chemical COMPOUNDS

chemical
COMPOUNDS

NEIL SCHLAGER, JAYNE WEISBLATT, AND
DAVID E. NEWTON, *EDITORS*

Charles B. Montney, *Project Editor*

VOLUME  **2**

ETHYL ALCOHOL–
POLYSILOXANE

U·X·L
An imprint of Thomson Gale,
a part of The Thomson Corporation

THOMSON
———✳———™
GALE

Detroit • New York • San Francisco • San Diego • New Haven, Conn. • Waterville, Maine • London • Munich

Chemical Compounds

Neil Schlager, Jayne Weisblatt, and David E. Newton, Editors

Project Editor
Charles B. Montney

Editorial
Luann Brennan, Kathleen J. Edgar, Jennifer Greve, Madeline S. Harris, Melissa Sue Hill, Debra M. Kirby, Kristine Krapp, Elizabeth Manar, Kim McGrath, Paul Lewon, Heather Price, Lemma Shomali

Indexing Services
Barbara Koch

Imaging and Multimedia
Randy Bassett, Michael Logusz

Product Design
Kate Scheible

Composition
Evi Seoud, Mary Beth Trimper

Manufacturing
Wendy Blurton, Dorothy Maki

LIBRARY OF CONGRESS CATALOGING-IN-PUBLICATION DATA

Weisblatt, Jayne.
    Chemical compounds / Jayne Weisblatt ; Charles B. Montney, project editor.
      v. cm.
    Includes bibliographical references and indexes.
    Contents: v. 1. Acetaminophen through Dimethyl ketone -- v. 2. Ethyl acetate through Polypropylene -- v. 3. Polysiloxane through Zinc oxide.
    ISBN 1-4144-0150-7 (set : alk. paper) -- ISBN 1-4144-0451-4 (v. 1 : alk. paper) -- ISBN 1-4144-0452-2 (v. 2 : alk. paper) -- ISBN 1-4144-0453-0 (v. 3 : alk. paper)
  1. Chemicals. 2. Organic compounds. 3. Inorganic compounds. I. Montney, Charles B., 1962- II. Title.
    QD471.W45 2006
    540--dc22
                                          2005023636

This title is also available as an e-book
ISBN 1-4144-0467-0
Contact your Thomson Gale sales representative for ordering information.

Printed in China
10 9 8 7 6 5 4 3 2 1

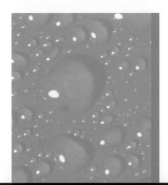

contents

Contents

volume 2 *Reader's Guide* xiii

Contents

Contents

Contents

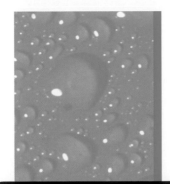

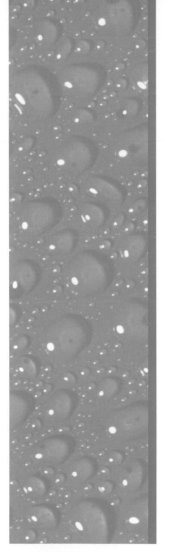

reader's guide

Water; sugar; nylon; vitamin C. These substances are all very different from each other. But they all share one property in common: They are all chemical compounds. A chemical compound consists of two or more chemical elements, joined to each other by a force known as a chemical bond.

This book describes 180 chemical compounds, some familiar to almost everyone, and some less commonly known. Each description includes some basic chemical and physical information about the compound, such as its chemical formula, other names by which the compound is known, and the molecular weight, melting point, freezing point, and solubility of the compound. Here are some things to know about each of these properties:

Other Names: Many chemical compounds have more than one name. Compounds that have been known for many centuries often have common names that may still be used in industry, the arts, or some other field. For example, muriatic acid is a very old name for the compound now called hydrochloric acid. The name remains in common use today. Marine acid and spirit of salt are other ancient names for hydrochloric acid, but they are seldom used in the modern world. All compounds have systematic names, names based on a set of rules devised by the International Union of Pure and Applied Chemistry (IUPAC). For example, the systematic name for the poisonous gas whose common name is mustard gas is 2,2'-dichlorodiethyl sulfide. When chemists talk about chemical compounds, they usually use only the official IUPAC name for a compound since that name leaves no doubt as to the substance about which they are talking. In some cases, a compound may have more than one official name, depending on the set of rules used in the naming process. For example, 1,1'-thiobis[2-chloroethane] is also an acceptable name for mustard gas. The "Other Names" section of each entry lists both the systematic (IUPAC) and common names for a compound.

Many compounds also have another kind of name, a brand name or trade name given to them by their manufacturers. For example, some trade names for the pain killer acetaminophen are Panadol™, Tylenol™, Aceta™, Genapap™, Tempra™, and Depacin™. The symbol next to each name means that the name is registered to the company that makes the compound. Trades names may be mentioned in the Overview or Uses sections of the entry for each compound.

Chemical Formula: A chemical formula is a set of symbols that tells the elements present in a compound and the relative numbers of each element. For example, the chemical formula for the compound carbon dioxide is CO_2. That formula tells that for every one carbon atom (C) in carbon dioxide there are two atoms of oxygen (O).

Chemists use different kinds of formulas to describe a compound. The simplest formula is a molecular formula. A molecular formula like CO_2 tells the kind and relative number of elements present in the compound. Another kind of formula is a structural formula. A structural formula provides one additional piece of information: The arrangement of elements in a compound. The structural formula for methanol (wood alcohol), for example, is CH_3OH. That formula shows that methanol consists of a carbon atom (C) to which are attached three hydrogen (H) atoms (CH_3). The carbon atom is also joined to an oxygen atom (O) which, in turn, is attached to a hydrogen atom (H).

Structural formulas can be written in a variety of ways. Another way to draw the structural formula for methanol, for example, is to show where individual bonds between atoms branch off other atoms in different directions. These structural formulas can be seen on the first page of nearly all entries in *Chemical Compounds*. In a third type of structural formula, the ball-and-stick formula, each element is represented by a ball of some size, shape, and/or color. The chemical bond that holds them together is represented by sticks. This can be represented on paper in a drawing that simulates a three-dimensional model, by computer software, or actually in three dimensions from a kit with balls and sticks.

All three kinds of structural formulas are given for each compound described in this book. The only exception is some very large compounds known as polymers that contain many hundreds or thousands of atoms. In such cases, the formulas given shown only one small segment of the compound.

Compound Type: Millions of chemical compounds exist. To make the study of these compounds easier, chemists divide them into a number of categories. Nearly all compounds can be classified as either organic or inorganic. Organic compounds contain the element carbon; inorganic compounds do not. A few important exceptions to that rule exist, as indicated in the description of such compounds.

Both organic and inorganic compounds can be further divided into more limited categories, sometimes called families of compounds. Some families of organic compounds are the hydrocarbons (made of carbon and hydrogen only), alcohols (containing the -OH group), and carboxylic acids (containing the -COOH groups). Many interesting and important organic compounds belong to the polymer family. Polymers consist of very large molecules in which a single small unit (called the monomer) is repeated hundreds or thousands of times over. Some polymers are made from two or, rarely, three monomers joined to each other in long chains.

Most inorganic compounds can be classified into one of four major groups. Those groups are the acids (all of which contain at least one hydrogen (H) atom), bases (which all have a hydroxide (OH) group), oxides (which all have an oxygen (O)), and salts (which include almost everything else). A few organic and inorganic compounds described in this book do not easily fit into any of these families. They are classified simply as organic or inorganic.

Molecular Weight: The molecular weight of a compound is equal to the weight of all the elements of which it is made. The molecular weight of carbon dioxide (CO_2), for example, is equal to the atomic weight of carbon (12) plus two times the atomic weight of oxygen (2 x 16 = 32), or 44. Chemists have been studying atomic weights and molecular weights for a long time, and the molecular weights of most compounds are now known with a high degree of certainty. The molecular weights expressed in this book are taken from the *Handbook of Chemistry and Physics*, 86th edition, published in 2005. The Handbook is one of the oldest, most widely used, and most highly regarded reference books in chemistry.

Melting Point and Boiling Point: The melting point of a compound is the temperature at which it changes from a solid to a liquid. Its boiling point is the temperature at which it changes from a liquid to a gas. Most organic compounds have precise melting points and/or, sometimes, precise boiling points. This fact is used to identify organic compounds. Suppose a chemist finds that a certain unknown compound melts at exactly 16.5°C. Reference books show that only a small number of compounds melt at exactly that temperature (one of which is capryllic acid, responsible for the distinctive odor of some goats). This information helps the chemist identify the unknown compound.

Inorganic compounds usually do not have such precise melting points. In fact, they may melt over a range of temperatures (from 50°C to 55°C, for example) or sublime without melting. Sublimation is the process by which a substance changes from a solid to gas without going through the liquid phase. Other inorganic compounds decompose, or break apart, when heated and do not have a true melting point.

Researchers often find different melting points and boiling points for the same compound, depending on the reference book they use. The reason for this discrepancy is that many scientists have measured the melting points and boiling points of compounds. Those scientists do not always get the same result. So, it is difficult to know what the "true" or "most correct" value is for these properties. In this book, the melting points and boiling points stated are taken from the *Handbook of Chemistry and Physics*.

Some compounds, for a variety of reasons, have no specific melting or boiling point. The term "not applicable" is used to indicate this fact.

Solubility: The solubility of a compound is its tendency to dissolve in some (usually) liquid, such as water, alcohol, or acetone. Solubility is an important property because most chemical reactions occur only when the reactants (the substances reacting with each other) are dissolved. The most common solvent for inorganic compounds is water. The most common solvents for organic compounds are the so-called organic solvents, which include alcohol, ether, acetone, and benzene. The solubility section in the entry for each compound lists the solvents in which it will dissolve well (listed as "soluble"), to a slight extent ("slightly soluble"), or not at all ("insoluble").

Overview: The overview provides a general introduction to the compound, with a pronunciation of its name, a brief history of its discovery and/or use, and other general information.

How It Is Made: This section explains how the compound is extracted from the earth or from natural materials and/or how it is made synthetically (artificially). Some production methods are difficult to describe because they include reactants (beginning compounds) with difficult chemical names not familiar to most people with little or no background in chemistry. Readers with a special interest in the synthesis (artificial production) of these compounds should consult their local librarian or a chemistry teacher at a local high school or college for references that contain more information on the process in question. The For Further Information section may also contain this information.

Interesting Facts This section contains facts and tidbits of information about compounds that may not be essential to a chemist, an inventor, or some other scientific specialist, but may be of interest to the general reader.

Common Uses and Potential Hazards Chemical compounds are often of greatest interest because of the way they can be used in medicine, industry, or some other practical application. This section lists the most important uses of each compound described in the book.

All chemical compounds pose some risk to humans. One might think that water, sugar, and salt are the safest compounds in the world. But, of course, one can drown in water, become seriously overweight by eating too much sugar, and develop heart problems by using too much salt. The risk posed by a chemical compound really depends on a number of factors, one of the most important of which is the amount of the compound to which one is exposed. The safest rule to follow in dealing with chemical compounds is that they are ALL dangerous under some circumstances. One should always avoid spilling any chemical compound on the skin, inhaling its fumes, or swallowing any of the compound. If an accident of this kind occurs, one should seek professional medical advice immediately. This book is not a substitute for prompt first aid properly applied.

Having said all that, some compounds do pose more serious health threats than others, and some individuals are at greater risks than others. Those special health risks are mentioned toward the end of the "Common Uses and Potential Hazards" section of each entry.

For Further Information As the name suggests, this section provides ideas for books, articles, and Internet sources that provide additional information on the chemical compound listed.

ADDED FEATURES

Chemical Compounds contains several features to help answer questions related to compounds, their properties, and their uses.

- The book contains three appendixes: a list by formula, list by element contained in compounds, and list by type of compound.

- Each entry contains up to two illustrations to show the relationship of the atoms in a compound to each other, one a black and white structural formula, and one a color ball-and-stick model of a molecule or portion of a molecule of the compound.

- A chronology and timeline in each volume locates significant dates in the development of chemical compounds with other historical events.

- "For Further Information," a list of useful books, periodicals, and websites, provides links to further learning opportunities.

- The comprehensive index, which appears in each volume, quickly points readers to compounds, people, and events mentioned throughout *Chemical Compounds*.

ACKNOWLEDGMENTS

In compiling this reference, the editors have been fortunate in being able to rely upon the expertise and contributions of the following educators who served as advisors:

Ruth Mormon, Media Specialist, The Meadows School, Las Vegas, Nevada

Cathy Chauvette, Sherwood Regional Library, Alexandria, Virginia

Jan Sarratt, John E. Ewing Middle School, Gaffney, South Carolina

Rachel Badanowski, Southfield High School, Southfield, Michigan

The editors would also like to thank the artists of Publishers Resource Group, under the lead of Farley Pedini, for their fast and accurate work and grace under pressure.

COMMENTS AND SUGGESTIONS

We welcome your comments on *Chemical Compounds*. Please write: Editors, *Chemical Compounds*, U X L, 27500 Drake Rd., Farmington Hills, MI 48331; call toll-free 1-800-877-4253; fax, 248-699-8097; or send e-mail via http://www.gale.com.

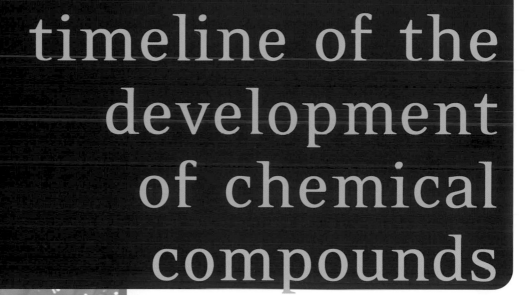

timeline of the development of chemical compounds

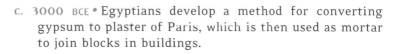

C. 3000 BCE • Egyptians develop a method for converting gypsum to plaster of Paris, which is then used as mortar to join blocks in buildings.

C. 2700 BCE • Chinese documents mention sodium chloride and the consumption of tea.

C. 1550 BCE • The analgesic properties of willow tree bark, from which salicylic acid comes, are described in Egyptian scrolls.

C. 1000 BCE • Ancient Egyptians use dried peppermint leaves.

800 BCE • Chinese and Arabic civilizations use borax for making glass and in jewelry work.

510 BCE • Persian emperor Darius makes the first recorded reference to sugar when he refers to the sugar cane growing on the banks of the Indus River.

184 BCE • Roman writer Cato the Elder describes a method of producing calcium oxide.

c. 1st century CE • Roman philosopher Pliny the Elder writes about a substance he calls hammoniacus sal, which appears to have been ammonium chloride.

1st century CE • The first recipes calling for the use of pectin to make jams and jellies are recorded.

c. 575 CE • The cultivation of the coffee tree begins in Africa.

659 • Cinnamaldehyde is described in the famous Chinese medical text, the *Tang Materia Medica.*

8th century • Arabian chemist Abu Musa Jabir ibn Hayyan, also known as Geber, writes about his work with several compounds, such as sodium chloride, sulfuric acid, nitric acid, citric acid, and acetic acid.

1242 • English natural philosopher Roger Bacon describes a method for making gunpowder.

Late 1200s • First mention of camphor by a Westerner occurs in the writings of Marco Polo.

4000 BCE Iron Age begins in Egypt.			8TH CENTURY BCE First recorded Olympic Games.	C. 6 BCE Jesus of Nazareth is born.	622 Mohammed's flight from Mecca to Medina.
4000 BCE	3000 BCE	2000 BCE	1000 BCE	1 CE	500

1300s • Potassium sulfate becomes known to alchemists.

1500s • Spanish explorers bring vanilla to Europe from South and Central America, where it had already been used to flavor food.

1603 • Flemish chemist Jan Baptista van Helmont isolates a new gas produced during the combustion of wood, which is eventually called carbon dioxide.

1608 • Potash is one of the first chemicals to be exported by American colonists, with shipments leaving Jamestown, Virginia.

1610 • French alchemist Jean Béguin prepares acetone.

1620 • Flemish physician and alchemist Jan Baptista van Helmont first discovers nitric oxide.

1625 • German chemist Johann Rudolf Glauber is believed to have been the first to produce hydrogen chloride in a reasonably pure form. Later he is first to make ammonium nitrate artificially.

1695 • The term *Epsom salts* is introduced by British naturalist Nehemiah Grew, who names the compound after the spring waters near Epsom, England, from which it was often extracted.

	1215		1492		
1096–1099	Magna Carta accepted by King		Christopher Columbus sails	1620 Pilgrims land at	
First Crusade	John of England.		to the Americas.	Plymouth, Mass.	

| 1000 | 1200 | 1400 | 1500 | 1600 | 1700 |

1700 • German chemist Georg Ernst Stahl extracts acetic acid from vinegar by distillation.

1702 • German chemist Wilhelm Homberg is believed to be the first person to prepare boric acid in Europe.

1720s • German chemist Johann Schulze makes discoveries that lead to using silver nitrate in printing and photography.

1746 • The first commercially successful method for making sulfuric acid is developed.

1747 • German chemist Andreas Sigismund Marggraf isolates a sweet substance from raisins that comes to be known as glucose.

1753 • James Lind reports that citrus fruits are the most effective means of preventing scurvy.

1769 • Oxalic acid is first isolated by German chemist Johann Christian Wiegleb.

1770s • British chemist Joseph Priestly does pioneering work with the compounds carbon dioxide, carbon monoxide, hydrogen chloride, and nitrous oxide, among others.

1770s • Swedish chemist Karl Wilhelm Scheele discovers and works with phosphoric acid, glycerol, lactic acid, and potassium bitartrate.

1726
Czar Peter the Great
of Russia dies.

1754
French and Indian
War begins in
North America.

| 1700 | 1710 | 1720 | 1730 | 1740 | 1750 |

1773 • French chemist Hilaire Marin Rouelle identifies urea as a component of urine.

Late 1700s • Commercial production of sodium bicarbonate as baking soda begins.

1776 • Carbon monoxide is first prepared synthetically by French chemist Joseph Marie François de Lassone, although he mistakenly identifies it as hydrogen.

1790 • The first patent ever issued in the United States is awarded to Samuel Hopkins for a new and better way of making pearl ash.

1794 • Ethylene is first prepared by a group of Dutch chemists.

Early 1800s • Silver iodide is first used in photography by French experimenter Louis Daguerre.

1817-1821 • French chemists Joseph Bienaimé Caventou and Pierre Joseph Pelletier successfully extract caffeine, quinine, strychnine, brucine, chinchonine, and chlorophyll from a variety of plants.

1817 • Irish pharmacist Sir James Murray uses magnesium hydroxide in water to treat stomach and other ailments. The compound is eventually called milk of magnesia.

				1811-1812	
1776			1793	Three severe	
The U.S. Declaration			Cotton gin is	earthquakes	1812.
of Independence is		1789	invented by	occur near New	War of 1812
signed.		French Revolution	Eli Whitney.	Madrid, Missouri.	begins.

1760	1770	1780	1790	1800	1810

1818 • Hydrogen peroxide is discovered by French chemist Louis Jacques Thénard.

1819 • French naturalist Henri Braconnot discovers cellulose.

1825 • British chemist and physicist Michael Faraday discovers "bicarburet of hydrogen," which is later called benzene.

1830 • Peregrine Phillips, a British vinegar merchant from England, develops the contact process for making sulfuric acid. In the early 21st century it is still the most common way to make sulfuric acid.

1831 • Chloroform is discovered almost simultaneously by American, French, and German chemists. Its use as an anesthetic is discovered in 1847.

1831 • Beta-carotene is first isolated by German chemist Heinrich Wilhelm Ferdinand Wackenroder.

1834 • Cellulose is first isolated and analyzed by French botanist Anselme Payen.

1835 • Polyvinyl chloride is first discovered accidentally by French physicist and chemist Henry Victor Regnault. PVC is rediscovered (again accidentally) in 1926.

1819
U.S. acquires
Florida from
Spain.

1820
The Missouri
Compromise
is enacted.

1823
U.S. president
James Monroe
proclaims the
Monroe Doctrine.

1831
Cyrus McCormick's
reaper is introduced.

1810   1820   1830

1836 • British chemist Edmund Davy discovers acetylene.

1838 • French chemist Pierre Joseph Pelletier discovers toluene.

1839 • German-born French chemist Henri Victor Regnault first prepares carbon tetrachloride.

1839 • German druggist Eduard Simon discovers styrene in petroleum.

1845 • Swiss-German chemist Christian Friedrich Schönbein discovers cellulose nitrate.

1846 • Americans Austin Church and John Dwight form a company to make and sell sodium bicarbonate. The product will become known as Arm & Hammer® baking soda.

Mid 1800s • Hydrogen peroxide is first used commercially–primarily to bleach hats.

1850s • Oil is first discovered in the United States in western Pennsylvania.

| 1846 Mexican-American War begins. | 1847 Gold discovered in California. | 1850 Levi Strauss manufactures his first pair of jeans. | 1858 Lincoln debates Douglas in Illinois senate campaign. |

| 1840 | 1850 | 1860 |

1853 • French chemist Charles Frederick Gerhardt develops a method for reacting salicylic acid (the active ingredient in salicin) with acetic acid to make the first primitive form of aspirin.

1859 • Ethylene glycol and ethylene oxide are first prepared by French chemist Charles Adolphe Wurtz.

1860s • Swedish chemist Alfred Nobel develops a process for manufacturing nitroglycerin on a large scale.

1863 • TNT is discovered by German chemist Joseph Wilbrand, although the compound is not recognized as an explosive until nearly 30 years later.

1865 • The use of carbolic acid as an antiseptic is first suggested by Sir Joseph Lister.

1865 • German botanist Julius von Sachs demonstrates that chlorophyll is responsible for photosynthetic reactions that take place within the cells of leaves.

1870 • American chemist Robert Augustus Chesebrough extracts and purifies petrolatum from petroleum and begins manufacturing it, eventually using the name Vaseline™.

1873 • German chemist Harmon Northrop Morse rediscovers and synthesizes acetaminophen. It had been discovered originally in 1852, but at the time it was ignored.

1861 American Civil War starts.	1866 Mendel discovers laws of heredity.	1869 Dmitri Mendeleev formulates the periodic law.	1876 Alexander Graham Bell patents the telephone.	1884 Worldwide system of standard time is adopted.

1860   1870   1880

1879 • Saccharin, the first artificial sweetener discovered, is synthesized accidentally by Johns Hopkins researchers Constantine Fahlberg and Ira Remsen.

1879 • Riboflavin is first observed by British chemist Alexander Wynter Blyth.

1883 • Copper(I) oxide is the first substance found to have semiconducting properties.

1886 • American chemist Charles Martin Hall invents a method for making aluminum metal from aluminum oxide, which drastically cuts the price of aluminum.

1889 • French physiologist Charles E. Brown Séquard performs early experiments on the effects of testosterone.

1890s • Commercial production of perchlorates begins.

1890s-early 1900s • British chemists Charles Frederick Cross, Edward John Bevan, and Clayton Beadle identify the compound now known as cellulose. They also develop rayon.

Late 1890s • Artificial methods of the production of pure vanillin are developed.

1901 • The effects of fluorides in preventing tooth decay are first observed.

1888
George Eastman
introduces the
Kodak camera.

1896
Henry Ford
assembles the
first motor car.

1901
The first Nobel
prizes are awarded.

1890   1900   1910

1904 • German physicist Wilhelm Hallwachs discovers that a combination of copper metal and copper(I) oxide displays the photoelectric effect.

1910 • The first plant for the manufacture of rayon in the United States is built. By 1925 rayon becomes more popular than silk.

1910 • Japanese scientist Suzuki Umetaro discovers thiamine.

1912 • Nicotinic acid is first isolated by Polish-American biochemist Casimir Funk.

1914-1917 • Ethylene glycol is manufactured for use in World War I in explosives and as a solvent.

1914-1918 • A shortage of sugar during World War I leads to the reintroduction of saccharin to sweeten food. Saccharin production would boom again during World War II.

Late 1910s • Mustard gas is first used in war. Later it is found to be effective in treating cancer in experimental animals.

1922 • Vitamin E is discovered by two scientists at the University of California at Berkeley.

1905
Einstein
formulates
the theory of
relativity.

1912
The *Titanic*
sinks on her
maiden voyage.

1914-1918
World War I

1919
First man-made
nuclear reaction
occurs.

1922
Soviet Union
comes into
existance.

1910   1915   1920

1928 • Polymethylmethacrylate (PMMA) is first synthesized in the laboratories of the German chemical firm Röhm and Haas.

1928 • Scottish bacteriologist Alexander Fleming accidentally discovers penicillin.

1930 • DuPont begins manufacturing dichlorodifluoromethane under the name Freon®.

1930 • A polymer based on styrene is produced by researchers at the German chemical firm I. G. Farben. Polystyrene comes to the United States in 1937.

1930s • Properties and methods of synthesizing many vitamins, such as riboflavin, thiamine, niacin, ascorbic acid, pyridoxine, alpha-tocopherol, and retinol, are developed.

Early 1930s • SBS is first developed by German chemists Walter Bock and Eduard Tschunkur.

1933 • British chemists Reginald Gibson and Eric Fawcett accidentally re-discover polyethylene, which was first discovered in 1889.

1935 • Nylon is invented by Wallace Carothers. It is used in consumer products within three years.

1935 • German chemist Adolf Friedrich Johann Butenandt synthesizes testosterone.

1927
Charles Lindbergh
flies solo across the
Atlantic Ocean.

1929
The Great
Depression
begins

1925   1930   1935

1937 • German forensic scientist Walter Specht discovers that blood can act as the catalyst needed to produce chemiluminescence with luminol, a compound discovered in the late 1800s.

1937 • Plexiglas® (made from polymethylmethacrylate) is exhibited at the World's Trade Fair in Paris.

1937 • The basic process for making polyurethanes is first developed by German chemist Otto Bayer.

1937 • The cyclamate family of compounds is discovered by Michael Sveda, a graduate student at the University of Illinois.

1938 • Polytetrafluoroethylene is invented by Roy J. Plunkett by accident at DuPont's Jackson Laboratory.

1939 • Swiss chemist Paul Hermann Müller finds that DDT is very effective as an insecticide, which makes it useful in preventing infectious diseases such as malaria.

1939-1945 • During World War II, the U.S. military finds a number of uses for nylon, polyurethanes, polystyrene, percholorates, and silica gel.

Early 1940s • Penicillin is first produced for human use and is valuable in saving the lives of soldiers wounded in World War II.

	1939	1941 First regular television broadcasts begin.	1942 Irving Berlin writes song "White Christmas."		1945 U.S. drops two atomic bombs on Japan.	1946 First "baby boomers" are born.
	World War II begins.					

1935	1940	1945

1940s • A research chemist at the General Electric Company, E. G. Rochow, finds an efficient way of making organosiloxanes in large quantities.

1941 • Folic acid is isolated and identified by American researcher Henry K. Mitchell.

1941 • The first polyurethane adhesive, for the joining of rubber and glass, is made.

1942 • American researchers Harry Coover and Fred Joiner discover cyanoacrylate.

1946 • DEET is patented by the U.S. Army for use on military personnel working in insect-infested areas. It is made available to the public in 1957.

1947 • On April 16, an ammonium nitrate explosion in Texas City, Texas, becomes the worst industrial accident in U.S.

1950s • Earliest reports surface about athletes using testosterone to enhance their sports performance.

1950s • A stretchable material made of polyurethane, called spandex, is introduced.

1951 • Phillips Petroleum Company begins selling polypropylene under the trade name of Marlex®.

1953 • Polycarbonate, polyethylene, and synthetic rubber are developed.

| 1950 Korean War begins. | 1953 Molecular structure of DNA is discovered. | 1955 Jonas Salk invents the polio vaccine. | 1958 First satellite broadcasts occur. |

1950   1955   1960

1955 • Proctor & Gamble releases the first toothpaste containing stannous fluoride, Crest®.

Mid 1950s • Wham-O creates the hula-hoop—a ring of plastic that is made with low-grade polyethylene.

1956 • British chemist Dorothy Hodgkin determines the chemical structure of cyanocobalamin.

1958 • Scientist W. Barnes of the chemical firm T. & H. Smith in Edinburgh, Scotland, discovers denatonium benzoate.

Early 1960s • Ibuprofen is developed by researchers at the Boots Company, a British drug manufacturer.

1960s • Triclosan becomes a common ingredient in soaps and other cleaning projects.

1960s • MTBE is first synthesized by researchers at the Atlantic Richfield Corporation as an additive designed to increase the fuel efficiency of gasoline.

1962 • Amoxicillin is discovered by researchers at the Beecham pharmaceutical laboratories.

1965 • Aspartame is discovered accidentally by James M. Schlatter.

1963
U.S. President
John F. Kennedy
is assassinated.

1975
Vietnam War
ends.

1950   1960   1970

1980s • A ceramic form of copper(I) oxide is found to have superconducting properties at temperatures higher than previously known superconductors.

1980s • Polycarbonate bottles begin to replace the more cumbersome and breakable glass bottles.

1987 • Procter & Gamble seeks FDA approval of sucrose polyester. Ten years pass before the FDA grants that approval.

1994 • The U.S. Food and Drug Administration approves the sale of naproxen as an over-the-counter medication.

1995 • On April 19, American citizens Timothy McVeigh and Terry Nichols use a truckload of ammonium nitrate and other materials to blow up the Alfred P. Murrah Federal Building in Oklahoma City, Oklahoma.

Early 2000s • Some 350,000 propane-powered vehicles exist in the United States and about 4 million are used worldwide.

2004 • The leading chemical compound manufactured in the United States is sulfuric acid, with 37,515,000 metric tons (41,266,000 short tons) produced. Next is ethylene, with about 26.7 million metric tons (29.4 million short tons) produced.

1981
First personal computers become available.

1989
The oil tanker *Exxon Valdez* sinks off Alaska.

1991
Soviet Union is dissolved.

2001
World Trade Center in New York City is destroyed.

1980   1990   2000   2006

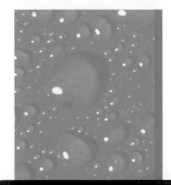

words to know

ACETYL The organic group of acetic acid.

ADHESIVE A substance used to bond two surfaces together.

ALCHEMY An ancient field of study from which the modern science of chemistry evolved.

ALKALI A chemical base that can combine with an acid to produce a salt.

ALKALINE A substance that has a pH higher than 7.

ALKALOID An organic base that contains the element nitrogen.

ALKANE A type of hydrocarbon that has no double bonds because it contains the maximum possible number of hydrogen

ALKENE A kind of hydrocarbon with at least one double bond between carbons.

ALKYL GROUP A chemical group containing hydrogen and carbon atoms.

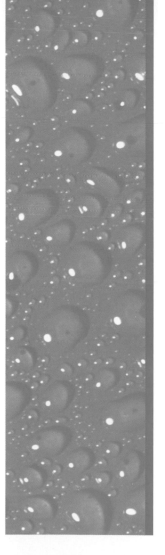

ALLOTROPE A form of an element that is different from its typical form, with a different chemical bond structure between atoms.

AMIDE An organic compound that includes the CON group bound to hydrogen.

AMINO ACID An organic compound that contains at least one carboxyl group (-COOH) and one amino group (-NH$_2$). They are the building blocks of which proteins are made.

ANALGESIC A substance the relieves pain.

ANHYDROUS Free from water and especially water that is chemically combined in a crystalline substance.

ANION A negatively charged ion.

ANODE The electrode in a battery in which electrons are lost (oxidized).

AROMATIC COMPOUND A compound whose chemical structure is based on that of benzene (C$_6$H$_6$).

BIODEGRADABLE Something that can be easily broken down by the action of bacteria.

BLOCK COPOLYMER A polymer composed of two or more different polymers, each of which clumps in blocks of identical molecules.

BORATE A salt that contains boron.

BRINE Salt water; water with a large amount of salt dissolved in it, such as seawater or water used to pickle vegetables.

BYPRODUCT A product that is made while making something else.

CARBOHYDRATES Organic compounds composed of carbon, oxygen, and hydrogen, which are used by the body as food.

CARBOXYL GROUPS Groups of atoms consisting of a carbon atom double bonded to an oxygen atom and single bonded to a hydroxyl (-OH) group (–COOH).

CARCINOGEN A substance that causes cancer in humans or other animals.

CATALYST A material that increases the rate of a chemical reaction without undergoing any change in its own chemical structure.

CATHODE The electrode in a battery through which electrons enter the fuel cell.

CATION A positively charged ion.

CAUSTIC Capable of burning or eating away, usually by the action of chemical reactions.

CENTRIFUGE A device that separates substances that have different densities by using centrifugal force.

CHELATE A chemical compound that is in the form of a ring. It usually contains one metal ion attached to a minimum of two nonmetal ions by coordinate bonds.

CHEMILUMINESCENCE Light produced by a chemical reaction.

CHIRAL A molecule with different left-handed and right-handed forms; not mirror symmetric.

CHLOROFLUOROCARBONS (CFCS) A family of chemicals made up of carbon, chlorine, and fluorine. CFCs were used as a refrigerant and propellant before they were banned for fear that they were destroying the ozone layer.

CHROMATOGRAPHY A process by which a mixture of substances passes through a column consisting of some material that causes the individual components in the mixture to separate from each other.

COAGULATE To make a liquid become a semisolid.

COENZYME A chemical compound that works along with an enzyme to increase the rate at which chemical reactions take place.

COMPOUND A substance formed of two or more elements that are chemically combined.

COPOLYMER A polymer made from more than one type of monomer.

COVALENT COMPOUND A compound in which the atoms are bonded to each other by sharing electrons.

CROSS-LINKED Polymer chains that are linked together to create a chemical bond.

CRYOGENICS The study of substances at very low temperatures, using substances such as liquefied hydrogen or liquefied helium.

DECOMPOSE To break a substance down into its most basic elements.

DENATURED To be made not fit for drinking.

DERIVATIVE Something gotten or received from another source.

DESICCANT Chemical agent that absorbs or adsorbs moisture.

DIATOMIC Composed of two atoms.

DISACCHARIDE A compound formed by the joining of two sugar molecules.

DISPERSANT A substance that keeps another substance from clumping together or becoming lumpy.

DISTILLATION A process of separating liquid by heating it and then condensing its vapor.

ELASTOMER A polymer known for its flexibility and elastic qualities; a type of rubber.

ELECTRODE A conductor through which an electric current flows.

ELECTROLYSIS A process in which an electric current is used to bring about chemical changes.

ELECTROLYTE A substance which, when dissolved in water, will conduct an electric current.

EMULSIFIER A substance that combines two other substances together that do not usually mix together and ensures they are spread evenly.

ESTER A compound formed by the reaction between an acid and an alcohol.

EXOTHERMIC Accompanied by the freeing of heat.

FAT An ester formed in the reaction between glycerol $(C_3H_5(OH)_3)$ and a fatty acid, an organic acid with more than eight carbon atoms.

FERROUS Containing or made from iron.

FLOCCULANT A type of polymer, or large man-made particle, that is created by a repetitive chain of atoms.

FLUOROCARBON A chemical compound that contains carbon and fluorine, such as chlorofluorocarbon.

FOSSIL FUEL A fuel, such as petroleum, natural gas, or coal, formed from the compression of plant and animal matter underground millions of years ago.

FREE RADICAL An atom or group of atoms with a single unpaired electron that can damage healthy cells in the body.

G/MOL Grams per mole: a measure of molar mass that indicates the amount of the compound that is found in a mole of the compound. The molecular weight (also known as relative molecular mass) is the same number as the molar mass, but the g/mol unit is not used.

GLOBAL WARMING The increase in the average global temperature.

GLUCONEOGENESIS The production of glucose from non-carbohydrate sources, such as proteins and fats.

GLYCOGEN A carbohydrate that is stored in the liver and muscles, which breaks down into glucose.

GREENHOUSE EFFECT The increase of the average global temperature due to the trapping of heat in the atmosphere.

HARD WATER Water with a high mineral content that does not lather easily.

HELIX A spiral; a common shape for protein molecules.

HORMONE A chemical that delivers messages from one cell or organ to another.

HYDRATE A chemical compound formed when one or more molecules of water is added physically to the molecule of some other substance.

HYDROCARBON A chemical compound consisting of only carbon and hydrogen, such as fossil fuels.

HYDROGENATION A chemical reaction of a substance with molecular hydrogen, usually in the presence of a catalyst.

HYDROLYSIS The process by which a compound reacts with water to form two new compounds.

INCOMPLETE COMBUSTION Combustion that occurs in such a way that fuel is not completely oxidized.

INERT A substance that is chemically inactive.

INORGANIC Relating to or obtained from nonliving things.

ION An atom or molecule with an electrical charge, either positive or negative.

IONIC BOND A force that attracts and holds positive and negative ions together.

IONIC COMPOUND A compound that is composed of positive and negative ions, so that the total charge of the positive ions is balanced by the total charge of the negative ions.

ISOMER Two or more forms of a chemical compound with the same molecular formula, but different structural formulas and different chemical and physical properties.

ISOTOPE A form of an element with the usual number of protons in the nucleus but more or less than the usual number of electrons.

KREBS CYCLE A series of chemical reactions in the body that form part of the pathway by which cells break down carbohydrates, fats, and proteins for energy. Also called the citric acid cycle.

LATENT Lying hidden or undeveloped.

LEACH Passing a liquid through something else in order to dissolve minerals from it.

LEWIS ACID An acid that can accept two electrons and form a coordinate covalent bond.

LIPID An organic compound that is insoluble in water, but soluble in most organic solvents, such as alcohol, ether, and acetone.

MEGATON A unit of explosive force equal to one million metric tons of TNT.

METABOLISM All of the chemical reactions that occur in cells by which fats, carbohydrates, and other compounds are broken down to produce energy and the compounds needed to build new cells and tissues.

METALLURGY The science of working with metals and ores.

METHYLXANTHINE A family of chemicals including caffeine, theobromine, and theophylline, many of which are stimulants.

MINERALOGIST A scientist who studies minerals.

MISCIBLE Able to be mixed; especially applies to the mixing of one liquid with another.

MIXTURE A collection of two or more elements or compounds with no definite composition.

MONOMER A single molecule that can be strung together with like molecules to form a polymer.

MONOSACCHARIDE A simple sugar, made up of three to nine carbon atoms.

MORDANT A substance used in dyeing and printing that reacts chemically with both a dye and the material being dyed to help hold the dye permanently to the material.

NARCOTIC An addictive drug that relieves pain and causes drowsiness.

NEUROTRANSMITTER A chemical that relays signals along neurons.

NEUTRALIZE To make a substance neutral that is neither acidic or alkaline.

NITRATING AGENT A substance that turns other substances into nitrates, which are compounds containing NO_3.

NSAID Non-steroidal anti-inflammatory drug, a drug that can stop pain and prevent inflammation and fever but that is not a steroid and does not have the same side effects as steroids.

ORGANIC Relating to or obtained from living things. In chemistry, refers to compounds made of carbon combined with other elements.

OXIDANT A substance that causes oxidation of a compound by removing electrodes from the compound. Also known as oxidizing agent.

OXIDATION STATE The sum of negative and positive charges, which indirectly indicates the number of electrons accepted or donated in the bond between elements.

OXIDIZE To combine a substance with oxygen, or to remove hydrogen from a molecule using oxygen, or to remove electrons from a molecule.

PARTICULATE MATTER Tiny particles of pollutants suspended in the air.

PETROCHEMICALS Chemical compounds that form in rocks, such as petroleum and coal.

PH (POTENTIAL HYDROGEN) The acidity or alkalinity of a substance based on its concentration of hydrogen ions.

PHOSPHATE A compound that is a salt of phosphoric acid, which consists of phosphorus, oxygen, sometimes hydrogen, and another element or ion.

PHOTOELECTRIC EFFECT The emission of electrons by a substance, especially metal, when light falls on its surface.

PHOTOSYNTHESIS The process by which green plants and some other organisms using the energy in sunlight to convert carbon dioxide and water into carbohydrates and oxygen.

PHOTOVOLTAIC EFFECT A type of photoelectric effect where light is converted to electrical voltage by a substance.

PLASTICIZER A substance added to plastics to make them stronger and more flexible.

POLYAMIDE A polymer, such as nylon, containing recurrent amide groups linking segments of the polymer chain.

POLYMER A substance composed of very large molecules built up by linking small molecules over and over again.

POLYSACCHARIDE A very large molecule made of many thousands of simple sugar molecules joined to each other in long, complex chains.

PRECIPITATE A solid material that settles out of a solution, often as the result of a chemical reaction.

PRECURSOR A compound that gives rise to some other compound in a series of reactions.

PROPRIETARY Manufactured, sold, or known only by the owner of the item's patent.

PROSTAGLANDINS A group of potent hormone-like substances that are produced in various tissues of the body. Prostaglandins help with a wide range of physiological functions, such as control of blood pressure, contraction of smooth muscles, and modulation of inflammation.

PROTEIN A large, complex compound made of long chains of amino acids. Proteins have a number of essential functions in living organisms.

QUARRY An open pit, often big, that is used to obtain stone.

REAGENT A substance that is employed to react, measure, or detect with another substance.

REDUCTION A chemical reaction in which oxygen is removed from a substance or electrons are added to a substance.

REFRACTORY A material with a high melting point, resistant to melting, often used to line the interior of industrial furnaces.

RESIN A solid or semi-solid organic material that is used to make lacquers, adhesives, plastics, and many other clear substances.

S

SALT An ionic compound where the anion is derived from an acid.

SEMICONDUCTOR A material that has an electrical conductance between that of an insulator and a conductor. When charged with electricity or light, semiconductors change their state from nonconductive to conductive or vice versa.

SEQUESTERING AGENT A substance that binds to metals in water to prevent them from combining with other components in the water and forming compounds that could stain (sequestering agents are sometimes used in cleaning products). Also called a "chelating agent."

SILICATE A salt in which the anion contains both oxygen and silicon.

SOLUBLE Capable of being dissolved in a liquid such as water.

SOLUTION A mixture whose properties and uniform throughout the mixture sample.

SOLVENT A substance that is able to dissolve one or more other substances.

SUBLIME To go from solid to gaseous form without passing through a liquid phase.

SUPERCONDUCTIVITY A state in which a material loses all electrical resistance: once established, an electrical current will flow forever.

SUPERCOOLED WATER Water that remains in liquid form, even though its temperature is below o°C.

SUSPENSION A mixture of two substances that do not dissolve in each other.

SYNTHESIZE To produce a chemical by combining simpler chemicals.

TERATOGENIC Causing birth defects; comes from the Greek word "teratogenesis," meaning "monster-making."

THERMAL Involving the use of heat.

TOXIC Poisonous or acting like a poison.

TOXIN A poison, usually produced by microorganisms or by plants or animals.

TRAP A reservoir or area within Earth's crust made of non-porous rock that can contain liquids or gases, such as water, petroleum, and natural gas.

ULTRAVIOLET LIGHT Light that is shorter in wavelength than visible light and can fade paint finishes, fabrics, and other exposed surfaces.

VASODILATOR A chemical that makes blood vessels widen, reducing blood pressure.

VISCOUS Having a syrupy quality causing a material to flow slowly.

VITRIFICATION The process by which something is changed into glass or a glassy substance, usually by heat.

VOLATILE Able to turn to vapor easily at a relatively low temperature.

WATER OF HYDRATION Water that has combined with a compound by some physical means.

$$H_3C \overset{\overset{\displaystyle H}{|}}{\underset{\underset{\displaystyle OH}{|}}{C}} H$$

OTHER NAMES:
Ethanol; grain
alcohol; alcohol; ethyl
hydrate

FORMULA:
CH_3CH_2OH

ELEMENTS:
Carbon, hydrogen,
oxygen

COMPOUND TYPE:
Alcohol (organic)

STATE:
Liquid

MOLECULAR WEIGHT:
46.07 g/mol

MELTING POINT:
−114.14°C (−173.45°F)

BOILING POINT:
78.29°C (172.9°F)

SOLUBILITY:
Miscible with water,
ether, acetone, and
most common
organic solvents

KEY FACTS

Ethyl Alcohol

OVERVIEW

Ethyl alcohol (ETH-uhl AL-ko-hol) is a clear, colorless, flammable liquid with a sharp, burning taste and a pleasant, wine-like odor. It is one of the first chemical substances discovered and used by humans. Ceramic jugs apparently designed to hold beer have been dated to the Neolithic Period, about 10,000 BCE. Some scholars suggest that humans may have learned how to make beer and incorporated it into their daily diets even before they made and used bread. The making and use of wine is a clear theme in Egyptian pictographs dating to the fourth millennium BCE. There probably does not exist a human culture today in which alcohol consumption does not occur. Today, beverages with alcohol content ranging as low as two to five percent ("near beer" and beer) to as high as 50 percent (some forms of vodka) are known and consumed by humans. In spite of its widespread use as a beverage, ethyl alcohol has a number of commercial and industrial uses that account for more than 90 percent of all the compound produced in the United States.

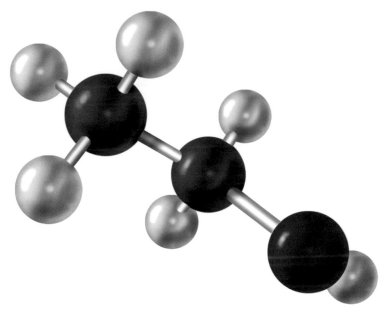

Ethyl alcohol. Red atom is oxygen; white atoms are hydrogen; and black atoms are carbon. PUBLISHERS RESOURCE GROUP

HOW IT IS MADE

Ethyl alcohol is made in one of two ways: naturally, through the process of fermentation, or synthetically, beginning with compounds found in petroleum. Until the beginning of World War II, more than 90 percent of all ethyl alcohol produced in the United States and other developed nations was made by fermentation. Waste syrup left over from the production of sugar from sugar cane was treated with enzymes at temperature of 20°C to 38°C (68°F to 100°F) for 28 to 72 hours. Under these conditions, about 90 percent of the syrup is converted to ethyl alcohol.

Over time, synthetic methods for the production of ethyl alcohol were developed. In one such method, ethylene (ethene; $CH_2=CH_2$) is treated with sulfuric acid and water to= obtain ethyl alcohol. That method was popular during the 1950s and 1960s. Then, a new method for making the compound was invented. In that process, ethylene and water are heated together at high temperatures [300°C to 400°C (570°F to 750°F)] and high pressures [1,000 pounds per square inch (6.9 megaPascals)] over a catalyst of phosphoric acid (H_3PO_4). The efficiency of this method is greater than

Interesting Facts

- All members of the alcohol family of organic compounds (such as methyl alcohol, ethyl alcohol, and isopropyl alcohol) are toxic to some extent. Only ethyl alcohol is safe to drink in relatively small quantities. Even then, a blood alcohol concentration of less than 5 percent can result in death.

- The concentration of alcohol in a beverage is often expressed as a "proof" number. The proof number of a beverage is twice that of its alcohol concentration. Thus, a beverage that is 80 proof has an alcohol concentration of 40 percent.

the older method, and there are fewer environmental consequences from making ethyl alcohol by this process.

As of 2003, about 94 percent of all ethyl alcohol was produced by fermentation. The remainder was produced by the phosphoric acid method.

COMMON USES AND POTENTIAL HAZARDS

In 2005, 10,500 million liters (2,790 million gallons) of ethyl alcohol were produced by fermentation methods. Of that amount, 92 percent was used as a fuel or an additive in fuels. Many experts suggest that consumers use a mixture of gasoline (90 percent) and ethyl alcohol (10 percent) called gasohol as a vehicle fuel because it burns more completely and releases fewer harmful byproducts to the environment. Although gasohol has not yet become very popular in the United States, it is widely used in some other parts of the world, most notably, in Brazil.

Of the remaining 8 percent of ethyl alcohol produced by fermentation, half was used in industrial operations, as a solvent or intermediary in the preparation of other chemical compounds; and half was used in the production of alcoholic beverages.

In 2005, about 650 million liters (170 million gallons) of ethyl alcohol were produced by the phosphoric acid method. Of that amount, 60 percent was used for industrial solvents in the manufacture of toiletries and cosmetics, coatings and inks, detergents and household cleaners, pharmaceuticals, and other products. The remaining 40 percent was used in the preparation of other chemical compounds, including ethyl acrylate, vinegar, ethylamines, ethyl acetate, glycol ethers, and miscellaneous materials.

Ethyl alcohol commonly occurs in one of three general forms. Absolute alcohol is ethyl alcohol that contains less than 1 percent impurities, such as water. Absolute alcohol is very difficult to make because ethyl alcohol will absorb water from the atmosphere or any other source that is available. The ethyl alcohol used in fuels and almost all industrial operations is a mixture of 95 percent ethyl alcohol and 5 percent water. Both absolute and 95 percent ethyl alcohol are extremely toxic. Ingestion of even very small amounts of either liquid has serious health effects that may include death.

The alcohol with which most people commonly come into contact is ethyl alcohol mixed with water in alcoholic beverages, such as beer, wine, gin, vodka, rum, or bourbon. In such beverages, the concentration of ethyl alcohol ranges from a few percent to 50 percent.

The effects produced by ethyl alcohol on the human body depend on the type of beverage consumed and the time taken for consumption. Drinking a 5-percent beer over an hour has a very different effect on the body than drinking a 50-percent vodka in five minutes.

Ethyl alcohol is a central nervous system depressant. After ingestion, it passes through a person's stomach and the small intestine, where it is absorbed rapidly into the bloodstream. It then travels throughout the body, interfering with the normal functioning of the nervous system and producing symptoms such as drowsiness, slurred speech, blurred vision, unsteady gait, impaired judgment, and reduced reaction time. With greater concentrations of alcohol in the blood, these symptoms may become more severe, resulting in coma and death.

Words to Know

CATALYST A material that increases the rate of a chemical reaction without undergoing any change in its own chemical structure.

MISCIBLE able to be mixed; especially applies to the mixing of one liquid with another.

SYNTHESIS A chemical reaction in which some desired chemical product is made from simple beginning chemicals, or reactants.

FOR FURTHER INFORMATION

"Alcohol." Erowid. http://www.erowid.org/chemicals/alcohol/alcohol.shtml (accessed on October 7, 2005).

"Alcohol: What You Don't Know Can Harm You." National Institute on Alcohol Abuse and Alcoholism. http://pubs.niaaa.nih.gov/publications/WhatUDontKnow_HTML /dontknow.htm (accessed on October 7, 2005).

Boggan, William. "Alcohol and You." http://chemcases.com/alcohol/ (accessed on October 7, 2005).

"Chemical of the Week: Ethanol." http://scifun.chem.wisc.edu/chemweek/ethanol/ethanol.html (accessed on October 7, 2005).

"Ethanol." http://www.ucc.ie/ucc/depts/chem/dolchem/html/comp/ethanol. html (accessed on October 7, 2005).

"How Alcohol Works." How Stuff Works. http://science.howstuffworks.com/alcohol.htm (accessed on October 7, 2005).

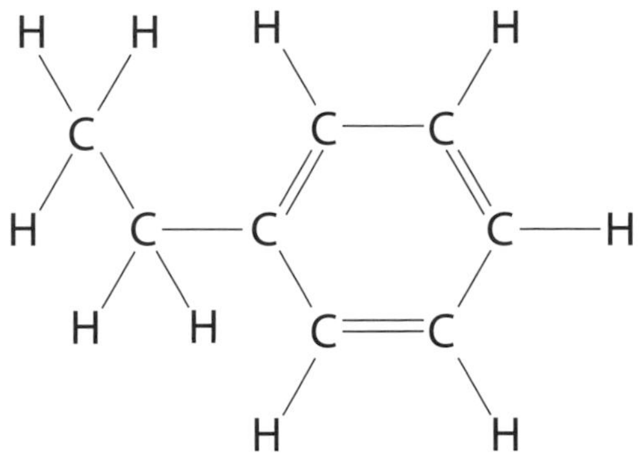

OTHER NAMES:
Phenylethane;
ethylbenzol

FORMULA:
$C_6H_5C_2H_5$

ELEMENTS:
Carbon, hydrogen

COMPOUND TYPE:
Aromatic
hydrocarbon
(organic)

STATE:
Liquid

MOLECULAR WEIGHT:
106.16 g/mol

MELTING POINT:
−94.96°C (−138.9°F)

BOILING POINT:
136.19°C (277.14°F)

SOLUBILITY:
Immiscible with
water; miscible with
ethyl alcohol and
methyl alcohol

KEY FACTS

Ethylbenzene

OVERVIEW

Ethylbenzene (eth-il-BEN-zeen) is a colorless flammable liquid with a pleasant aromatic odor. It is an aromatic hydrocarbon, that is, a compound consisting of carbon and hydrogen only with a molecular structure similar to that of benzene (C_6H_6). In 2004 it ranked fifteenth among chemicals produced in the United States. Its primary use is in the manufacture of another aromatic hydrocarbon, styrene ($C_6H_5CH=CH_2$), widely used to make a number of polymers, such as polystyrene, styrene-butadiene latex, SBR rubber, and ABS rubber.

HOW IT IS MADE

Ethylbenzene occurs to some extent as a component of petroleum. It can be extracted from petroleum by fractional distillation, the process by which individual components of petroleum are separated from each other by heating in a distilling tower. Ethylbenzene can also be made synthetically

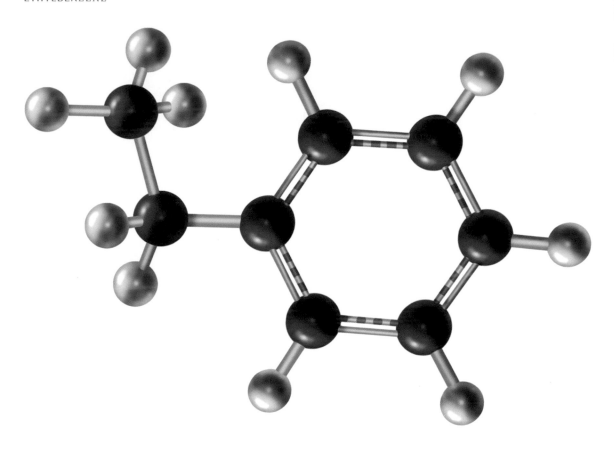

Ethylbenzene. Black atoms are carbon; white atoms are hydrogen. Bonds in the benzene ring are represented by the double striped sticks. White sticks show single bonds.
PUBLISHERS RESOURCE GROUP

by reacting benzene with ethene (ethylene; $CH_2=CH_2$) over a catalyst of aluminum chloride ($AlCl_3$): $C_6H_6 + CH_2=CH_2 \rightarrow C_6H_5C_2H_5$.

COMMON USES AND POTENTIAL HAZARDS

More than 99 percent of the ethylbenzene made is used for a single purpose—the production of styrene. Styrene is a very important industrial chemical, ranking seventeenth among all chemicals produced in the United States in 2004. It is used to make a number of important and popular polymers, the best known of which may be polystyrene. Much smaller amounts of ethylbenzene are used in solvents or as additives to a variety of products. Some products that contain ethylbenzene include synthetic rubber, gasoline and other fuels, paints and varnishes, inks, carpet glues, tobacco products, and insecticides.

Words to Know

CATALYST A material that increases the rate of a chemical reaction without undergoing any change in its own chemical structure.

IMMISCIBLE Does not mix with another liquid.

MISCIBLE Able to be mixed; especially applies to the mixing of one liquid with another.

POLYMER A compound consisting of very large molecules made of one or two small repeated units called monomers.

The concentration of ethylbenzene in consumer products is so low that it probably poses no threat to human health or to the environment. It may be a problem, however, when it leaks from industrial or chemical plants into the soil and becomes part of the groundwater. In such cases, it may be consumed by humans and other animals, or it may evaporate into the air, where it may be breathed by humans and other animals. Ethylbenzene has been found in about half (731 of 1467) of all the sites surveyed for pollutants by the U.S. Environmental Protection Agency, although concentrations are so low in most cases as to pose no threat to human health or the environment.

Ethylbenzene is an irritant to the skin, eyes, and respiratory system. In small quantities, it may cause dizziness, tightness in the chest, and burning of the eyes. In larger doses, it may cause narcotic effects, producing drowsiness and disorientation. The greatest safety concerns about ethylbenzene relate to its combustibility. Since it is more dense than air, it tends to settle to the ground and travel to any source of fire that may be near by.

FOR FURTHER INFORMATION

"Consumer Factsheet on: Ethylbenzene." U.S. Environmental Protection Agency.
http://www.epa.gov/safewater/contaminants/dw_contamfs/ethylben.html (accessed on December 22, 2005).

"Ethylbenzene." Australian Government. Department of the Environment and Heritage.
http://www.npi.gov.au/database/substance-info/profiles/40.html (accessed on December 22, 2005).

"ToxFAQs™ for Ethylbenzene." Agency for Toxic Substances and Disease Registry.
http://www.atsdr.cdc.gov/tfacts110.html (accessed on December 22, 2005).

See Also Polystyrene; Poly(Styrene-Butadiene-Styrene); Styrene

OTHER NAMES:
Ethene; bicarburetted
hydrogen; olefiant
gas

FORMULA:
CH₂=CH₂

ELEMENTS:
Carbon, hydrogen

COMPOUND TYPE:
Alkene; unsaturated
hydrocarbon
(organic)

STATE:
Gas

MOLECULAR WEIGHT:
28.05 g/mol

MELTING POINT:
−169.15°C (−272.47°F)

BOILING POINT:
−103.77°C
(−154.79°F)

SOLUBILITY:
Insoluble in water;
slightly soluble in
ethyl alcohol,
benzene, and acetone;
soluble in ether

KEY FACTS

Ethylene

OVERVIEW

Ethylene (ETH-ih-leen) is a colorless, flammable gas with a sweet odor and taste. It is the simplest alkene. Alkenes are hydrocarbons that contain one or more double bonds. Ethylene was first prepared in 1794 by a group of Dutch chemists including J. R. Deiman, A. Paets van Troostwyk, N. Bondt, and A. Lauwerenburgh. They treated ethanol (ethyl alcohol; C_2H_5OH) with concentrated sulfuric acid (H_2SO_4) and obtained ethylene, although they were incorrect in believing that the compound also contained oxygen.

Ethylene occurs naturally in petroleum and natural gas, but only to a very small percentage. It also occurs naturally in plants where it functions as a hormone and has a number of important effects on the growth and development of plants. These effects have been used for thousands of years, although the chemical mechanism involved was not understood. For example, the ancient Chinese are said to have burned incense in closed containers in order to facilitate the ripening of pears. Although they were certainly not

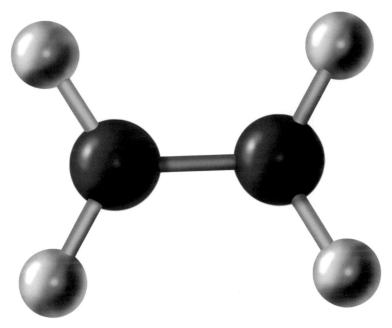

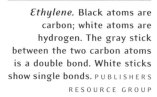

Ethylene. Black atoms are carbon; white atoms are hydrogen. The gray stick between the two carbon atoms is a double bond. White sticks show single bonds. PUBLISHERS RESOURCE GROUP

aware of the fact, the ripening effect was probably a result of ethylene gas released during combustion of the incense.

The hormonal effects of ethylene were first identified in 1901 by Russian chemist Dmitri N. Neljubow. For some time, scientists had observed that leaks from the gas lamps used to light city streets caused plants to grow old more rapidly than normal, often with strange changes in the structure of their leaves and stems. Neljubow was able to show that this effect was caused by ethylene in the lamp gas. Scientists have since learned a great deal about the production of ethylene in plants and the effects produced by the gas on the growth and development of those plants. For example, scientists discovered that ethylene is synthesized in germinating seeds, nodes of stems, and the tissue of ripening fruits. Its production is increased by flooding of the plant's roots; by drying of the soil; in response to environmental stress, such as attack by pests; by aging of the plant; and by inadequate amounts of minerals in the soil. They have also learned that ethylene increases the rate at which leaves and flowers age; promotes germination of seeds and the growth of root hairs; stimulates the ripening of fruit and flowers; and increases a plant's resistance to disease and physical damage.

HOW IT IS MADE

One reference on hydrocarbons lists more than 500 methods for making ethylene. From a practical standpoint, only a handful of those methods are important. The most common method of preparation involves the thermal or catalytic cracking of hydrocarbons. The term *cracking* refers to the process by which hydrocarbons from petroleum are broken down into simpler molecules. That process is usually accomplished by heating the petroleum for short periods of time at high temperatures (*thermal cracking*) or over a catalyst (*catalytic cracking*). In the cracking process, hydrocarbons with 10, 15, 20, or more carbon atoms are broken down to produce hydrocarbons with two, three, four, or some other small number of carbons. Ethylene is one of the usual products of cracking. It can be separated from the other products of cracking because it escapes from the reaction mixture as a gas.

Ethylene can also be produced from synthesis gas. Synthesis gas is the term used for various mixtures of gases produced when steam (with or without additional oxygen) is passed over hot coal. The steam and coal react to produce a rich mixture of hydrocarbons, a mixture that usually includes ethylene. Finally, ethylene can be produced in small quantities in the laboratory by the method first used by the Dutch chemists in 1794, namely by reacting ethanol with concentrated sulfuric acid.

COMMON USES AND POTENTIAL HAZARDS

In 2004, U.S. chemical manufacturers produced about 26.7 million metric tons (29.4 million short tons) of ethylene, making it the third most important chemical produced in the country in terms of volume. Over 90 percent of that ethylene was used for the production of other chemical compounds, the most important of which were polymers and related compounds, such as three major kinds of polyethylene (high density (HDP), low density (LDP), and linear low density polyethylene (LLDP)), polypropylene, ethylene dichloride, ethylene oxide, styrene, ethylbenzene, vinyl acetate, vinyl chloride, ethylene glycol, polystyrene, polyvinyl chloride (PVC), trichloroethylene, styrene-butadiene rubber (SBR), and ethyl alcohol.

Interesting Facts

- The highest rate of ethylene production yet measured in plants is that produced by the fading blossoms of the Vanda orchid, at 3.4 microliters per gram of flower per hour. The Vanda orchid is extensively used in the manufacture of Hawaiian leis.

- Air in rural areas typically contains about 5 parts per billion (ppb) of ethylene, while urban air generally contains about twenty times that amount.

- Ethylene is produced in plants in a complex series of reactions known as the Yang cycle (after its discoverer, S. F. Yang) that starts with the amino acid methionine.

- Production of ethylene in plants is significantly affected by a number of factors, such as temperature and presence of other gases (especially oxygen and carbon dioxide).

Smaller amounts of ethylene were used in a number of other chemical and industrial applications. These included:

- As a spray to accelerate the ripening of fruit;
- As a refrigerant;
- In oxygen-ethylene torches used in welding and metal-cutting operations;
- As a specialized anesthetic.

Ethylene poses a health hazard primarily because it is highly flammable and a serious explosive risk. It also acts as a narcotic at low concentrations, causing nausea, dizziness, headaches, and loss of muscular coordination. At higher concentrations, it acts as an anesthetic, causing loss of consciousness and insensitivity to pain and other stimuli. These effects tend to be of concern primarily to people who work directly with the gas. The amount of ethylene to which most people are exposed in their daily lives tends to be relatively low.

Words to Know

CATALYST A material that increases the rate of a chemical reaction without undergoing any change in its own chemical structure.

CRACKING The process by which larger hydrocarbons from petroleum are broken down into simpler molecules.

POLYMER : A compound consisting of very large molecules made of one or two small repeated units called monomers.

FOR FURTHER INFORMATION

"Ethene (ethylene): Properties, Production & Uses." Aus-e-tute. http://www.ausetute.com.au/ethene.html (accessed on December 27, 2005).

"Ethylene." Laboratory of Postharvest Physiology and Technology, Seoul National University. http://plaza.snu.ac.kr/~postharv/p3ethylene.pdf (accessed on December 27, 2005).

"Ethylene." National Safety Council. http://www.nsc.org/library/chemical/Ethylene.htm (accessed on December 27, 2005).

See Also Ethyl Alcohol; Ethylene Glycol; Ethylene Oxide; Ethylbenzene; Polyethylene; Polypropylene; Polystyrene; Polyvinyl Chloride

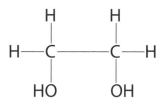

Ethylene Glycol

OVERVIEW

Ethylene glycol (ETH-uh-leen GLYE-kol) is clear, color-less, syrupy liquid with a sweet taste. One should not attempt to confirm the compound's taste, however, as it is toxic. In recent years, more than 4 billion kilograms (9 billion pounds) of ethylene glycol has been produced in the United States annually. The compound is used primarily as an antifreeze and in the manufacture of a number of important chemical compounds, including polyester fibers, films, bottles, resins, and other materials.

Ethylene glycol was first prepared in 1859 by the French chemist Charles Adolphe Wurtz (1817-1884). Wurtz's discovery did not find an application, however, until the early twentieth century, when the compound was manufactured for use in World War I (1914-1918) in the manufacture of explosives and as a coolant. By the 1930s, a number of uses for the compound had been found, and the chemical industry began producing ethylene glycol in large quantities.

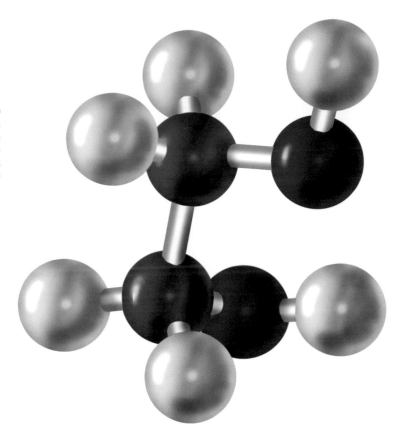

Ethylene glycol. Red atoms are oxygen; white atoms are hydrogen; and black atoms are carbon. PUBLISHERS RESOURCE GROUP

HOW IT IS MADE

The primary method of producing ethylene glycol involves the hydration of ethylene oxide, a ring compound consisting of two methylene ($-CH_2$) groups and one oxygen atom. Hydration is the process by which water is added to a compound. The hydration of ethylene oxide is conducted at a temperature of about 383°F (195°C) without a catalyst, or at about 50°C to 70°C (122°F to 158°F) with a catalyst, usually a strong acid, either process resulting in a yield of at least 90 percent of ethylene glycol.

Other methods of preparation are also available. For example, the compound can be produced directly from synthesis gas, a mixture of carbon monoxide and hydrogen; or by treating ethylene ($CH_2=CH_2$) with oxygen in an acetic acid solution using a catalyst of tellurium oxide or bromide ion.

Interesting Facts

- In 1996, 60 children died in Haiti of ethylene glycol poisoning after drinking cough syrup made in China that had accidentally been contaminated with the compound.

COMMON USES AND POTENTIAL HAZARDS

One of the first major uses of ethylene glycol was as a radiator coolant in airplanes. The compound actually made possible a change in the design of airplanes. At one time, plain water was used as the coolant in airplane radiators. The faster the airplane flew, the greater the risk that its radiator would boil over. Adding ethylene glycol to the water raised the boiling point of the coolant and allowed airplanes to fly faster with smaller radiators. This change was especially useful in the construction of military airplanes used in combat.

Ethylene glycol is still used extensively as a coolant and antifreeze in cooling systems. It is also used as a deicing fluid for airport runways, cars, and boats. Brake fluids and shock-absorber fluids often contain ethylene glycol as protection against freezing. About 26 percent of all the ethylene glycol made in the United States is used for some kind of cooling or antifreeze application.

The largest single use of ethylene glycol today is in the manufacture of a plastic called polyethylene terephthalate (PET). PET's primary application is in the manufacture of plastic bottles, an application that accounts for about a third of all the ethylene glycol made in the United States. Large amounts of PET are also used in the manufacture of polyester fibers and films. Some additional uses of the compound include:

- As a humectant (a substance that attracts moisture) in keeping some food, tobacco, and industrial products dry;

- As a solvent in some paints and plastics;

Words to Know

CATALYST A material that increases the rate of a chemical reaction without undergoing any change in its own chemical structure.

HUMECTANT A substance that attracts moisture more easily than other substances.

MISCIBLE able to be mixed; especially applies to the mixing of one liquid with another.

- In the dyeing of leathers and textiles;

- In the manufacture of printing inks, wood stains, ink for ball-point pens, and adhesives;

- In the production of artificial smoke and fog for theatrical productions;

- As a stabilizer in the soybean-based foam sometimes used to extinguish industrial fires; and

- In the manufacture of specialized types of explosives.

Ethylene glycol poses a number of potential health and safety hazards. It is very flammable and highly toxic. Ingestion of the compound may cause nausea, vomiting, abdominal pain, weakness, convulsions, and cardiac problems. Higher doses can result in severe kidney damage that leads to death.

FOR FURTHER INFORMATION

"Ethylene Glycol." Environmental Protection Agency. http://www.epa.gov/ttn/atw/hlthef/ethy-gly.html (accessed on October 7, 2005).

"Ethylene Glycol." National Safety Council. http://www.nsc.org/library/chemical/Ethylen1.htm (accessed on October 7, 2005).

"Medical Management Guidelines (MMGs) for Ethylene Glycol." Agency for Toxic Substances and Disease Registry. http://www.atsdr.cdc.gov/MHMI/mmg96.html (accessed on October 7, 2005).

H   O   H
 \   / \   /
  C-------C
 /   \
H   H

OTHER NAMES:
Epoxyethane; oxirane

FORMULA:
$(CH_2)_2O$

ELEMENTS:
Carbon, hydrogen,
oxygen

COMPOUND TYPE:
Cyclic ether (organic)

STATE:
Gas

MOLECULAR WEIGHT:
44.05 g/mol

MELTING POINT:
−112.5°C (−170.5°F)

BOILING POINT:
10.6°C (51.1°F)

SOLUBILITY:
Soluble in water,
ethyl alcohol,
acetone, benzene,
and ether

KEY FACTS

Ethylene Oxide

OVERVIEW

Ethylene oxide (ETH-ih-leen OK-side) is a flammable, color-less gas with the odor of ether. The gas is a cyclic compound, consisting of a ring of two carbon atoms and one oxygen atom. Each carbon atom also has two hydrogen atoms attached to it. Ethylene oxide was first prepared in 1859 by French chemist Charles Adolphe Wurtz (1817-1884). Wurtz produced the compound by reacting ethylene chlorhydrin (2-chloroethanol; $ClCH_2CH_2OH$) with an inorganic base (such as sodium hydroxide; NaOH), a process that remained the principle method for preparing the gas for more than a century. After World War II (1939-1945), a method was discovered for the direct oxidation of ethylene gas that is more efficient than the chlorhydrin process.

Ethylene oxide is a very unstable compound that catches fire or explodes readily and must be handled with the greatest care. Nonetheless, it is an important industrial chemical and ranks nineteenth by volume among chemicals produced in the United States. Its primary use is in the manufacture of other organic compounds.

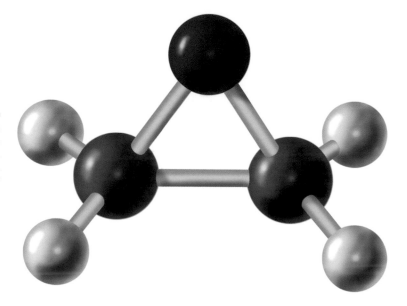

Ethylene Oxide. Black atoms are carbon; white atoms are hydrogen; red atom is oxygen. White sticks show single bonds.
PUBLISHERS RESOURCE GROUP

HOW IT IS MADE

The chlorhydrin process for making ethylene oxide has been replaced commercially by the direct oxidation of ethylene gas. Oxidation takes place at temperatures of 200°C to 300°C (392°F to 572°F) over a silver catalyst. The formula for this reaction is $2CH_2=CH_2 + O_2 \rightarrow 2CH_2CH_2O$. The yield produced by direct oxidation is slightly less than that produced by the chlohydrin process, but the amount of chlorine wasted by the latter method outweighs the slight difference in efficiency of production.

COMMON USES AND POTENTIAL HAZARDS

The largest single use of ethylene oxide is in the manufacture of ethylene glycol (CH_2OHCH_2OH), which itself is used as an antifreeze and raw material for the production of plastics. About 60 percent of all the ethylene oxide produced is used for this purpose. The compound is also used to make higher glycols, such as diethylene and triethylene glycol ($CH_2OHCH_2OCH_2CH_2OH$ and $CH_2OHCH_2OCH_2CH_2OCH_2CH_2OH$). The second most important application of ethylene oxide is in the synthesis of ethyoxylates and ethanolamines, substances used in the production of synthetic detergents. These substances act as surfactants in detergents—substances

Interesting Facts

- Ethylene oxide's instability is caused by its unusual three-atom ring structure. A ring with three atoms is a stressed arrangement that breaks apart with even moderate amounts of stress.

- A number of industrial accidents have been caused during the preparation, transportation, storage, and use of ethylene oxide. It is one of the most hazardous of the top twenty-five chemicals produced in the United States.

- In 2004, some 3.77 million metric tons (4.15 million short tons) of ethylene oxide were produced in the United States.

that reduce the surface tension between two materials and improve the lathering ability of a cleaner.

Ethylene oxide also has a number of other important industrial uses, although the quantity used for such purposes is small compared to the uses mentioned above. For example, it is used as a rocket propellant because of its tendency to decompose easily with the release of large amounts of energy. It is also used as a sterilizing medium, particularly for the sterilization of surgical instruments and consumer products, such as spices and cosmetics. The compound is also used as a demulsifier in the petroleum industry. A demulsifier is a material that aids in the separation of the components of complex mixtures, like those handled in the processing of petroleum. Ethylene oxide is also used as a fumigant, which is a gas used to kill insects and other pests.

A number of health hazards are associated with exposure to ethylene oxide. It is a skin, eye, and respiratory system irritant, causing symptoms such as dizziness, nausea, headache, convulsions, blistering of the skin, coughing, tightness of the chest, difficulty in breathing, and blurred vision. Long-term exposure to the gas may produce more serious health consequences, such as damage to the nervous system, muscular weakness, paralysis of the peripheral nerves, impaired thinking, loss of memory, and severe skin irritation. Ethylene

Words to Know

CATALYST A material that increases the rate of a chemical reaction without undergoing any change in its own chemical structure.

ETHER An organic compound in which one oxygen atom is bonded to two carbon atoms: - C · O · C. Ethylene oxide is the simplest cyclic ether.

oxide is believed to be responsible for spontaneous abortions, genetic damage, and the development of some types of cancer, primarily cancer of the blood (leukemia).

FOR FURTHER INFORMATION

Buckles, Carey, et al. *Ethyleneoxide.* 2nd ed. Celanese Ltd., The Dow Chemical Company, Shell Chemical Company, Sunoco, Inc., and Equistar Chemicals, LP. Also available online at http://www.ethyleneoxide.com/html/introduction.html (accessed January 3, 2006).

Environment Canada, Health Canada. *Ethylene Oxide.* Ottawa: Environment Canada, 2001.

"Ethylene Oxide." OSHA Fact Sheet. http://www.osha.gov/OshDoc/data_General_Facts/ethylene-oxide-factsheet.pdf (accessed on December 27, 2005).

"Ethylene Oxide Health and Safety Guide." IPCS International Programme on Chemical Safety. http://www.inchem.org/documents/hsg/hsg/hsg016.htm (accessed on December 27, 2005).

"ToxFAQs™ for Ethylene Oxide." Agency for Toxic Substances and Disease Registry. http://www.atsdr.cdc.gov/tfacts137.html (accessed on December 27, 2005).

See Also Ethylene; Ethylene Glycol

OTHER NAMES:
L-glutamic acid;
vitamin Bc; vitamin
B₉; vitamin M

FORMULA:
$C_{19}H_{19}N_7O_6$

ELEMENTS:
Carbon, hydrogen,
nitrogen, oxygen

COMPOUND TYPE:
Organic acid

STATE:
Solid

MOLECULAR WEIGHT:
441.40 g/mol

MELTING POINT:
Decomposes at 250°C
(480°F)

BOILING POINT:
Not applicable

SOLUBILITY:
Very slightly soluble
in water and methyl
alcohol; insoluble in
ethyl alcohol and
acetone

KEY FACTS

Folic Acid

OVERVIEW

Folic acid (FOH-lik AS-id) is a member of the B vitamin group, which is essential for the production of proteins and nucleic acids. In pure form, it is a tasteless, odorless, orange-to-yellow crystalline substance that is destroyed by heat or exposure to light. The compound occurs in three similar forms with comparable biological activity. Only one of its forms, l-pteroylglutamic acid, is made synthetically (in a laboratory). Folic acid is sometimes referred to in its ionic form as folate, which differs from folic acid only in the absence of a single hydrogen atom in its structure. The term folate is also used for a group of compounds structurally similar to folic acid.

Credit for the discovery of the nutritional significance of folic acid is often given to English medical researcher Lucy Wills (1888-1964). In the early 1920s, Wills discovered that anemia in pregnant women could be prevented if they included yeast in their diets. Anemia is a condition in which the blood contains too few red blood cells. Wills located

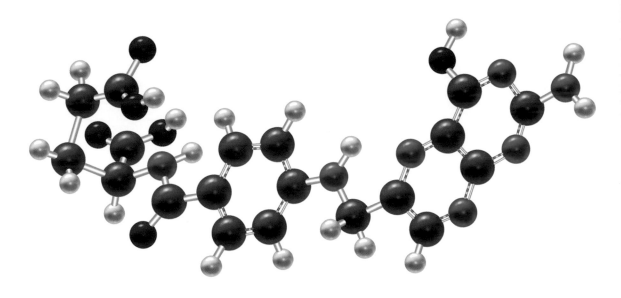

Folic Acid. Red atoms are oxygen; white atoms are hydrogen; black atoms are carbon; and blue atoms are nitrogen. Striped sticks indicate a benzene ring.
PUBLISHERS RESOURCE GROUP

a specific compound in yeast that produced this effect and called it the "Wills Factor." At about the same time, other research teams were discovering a similar compound that prevented anemia in monkeys, chicks, and other animals. One team, led by American researcher William J. Darby (1913-2001) called their anti-anemia factor "vitamin M," (for monkeys) and a third research team discovered a similar factor that prevented anemia in chicks and called it vitamin Bc (for chicks). Folic acid was finally isolated and identified in 1941 by American researcher Henry K. Mitchell (1917-), who suggested the modern name of folic acid for the compound. He chose the name because the compound was abundant in leafy vegetables and the Latin word for leaf is *folium.*

By the 1980s, scientists had produced evidence that the addition of folic acid in the diets of pregnant women can prevent birth defects such as spina bifida, a condition in which a baby's spinal column fails to close properly while developing inside the mother's womb. Researchers had learned how to fortify foods with folic acid in the 1970s, but manufacturers did not actually begin to do so until the late 1990s, when the U.S. government began requiring companies to supplement cereals, breads, and other grain-based products with the vitamin.

Interesting Facts

- The recommended daily dose of folic acid for adults is 400 micrograms per day. The best sources of the vitamin are fortified cereals and grain products, beef liver, black-eyed peas, spinach, avocado (raw), and eggs.

HOW IT IS MADE

The body produces some folic acid and obtains the remainder through food and dietary supplements. In the body, folic acid is produced by bacteria in the large intestine, absorbed in the small intestine, and stored in the liver. The l-pteroylglutamic acid form of folic acid is also produced synthetically.

COMMON USES AND POTENTIAL HAZARDS

The sole use of folic acid is as a nutrient in animal bodies. It is used in the synthesis of methionine, an amino acid used in the formation of proteins and nucleic acids. A deficiency of folic acid can produce various symptoms, including ulcers in the stomach and mouth, slowed growth, and diarrhea. It also results in a medical condition known as megaloblastic anemia, in which a person's body produces red blood cells that are larger than normal.

Adequate amounts of folic acid are especially important in fetal development, during the first eight weeks of life following fertilization. The compound is essential to promote normal development of the fetal nervous system. Folic acid deficiencies in the mother during this period may result in neural tube defects such as spina bifida or anencephaly, a condition in which the fetus' brain and skull fail to develop normally. The U.S. Centers for Disease Control and Prevention recommend that pregnant women take 600 micrograms of folic acid daily to avoid such problems.

Some evidence suggests that folic acid supplements may also reduce the risk of heart disease; strokes; and cervical and colon cancers. Folic acid is generally nontoxic, and side

Words to Know

SYNTHESIS A chemical reaction in which some desired chemical product is made from simple beginning chemicals, or reactants.

effects associated with its use are very rare. In unusual cases, allergic reactions to the compound have been reported.

FOR FURTHER INFORMATION

"Folate." PDRhealth.
http://www.pdrhealth.com/drug_info/nmdrugprofiles/nutsup-drugs/fol_0110.shtml (accessed on October 10, 2005).

Folate (Folacin, Folic Acid). Ohio State University Extension Fact Sheet.
http://ohioline.osu.edu/hyg-fact/5000/5553.html (accessed on October 10, 2005).

"Folic Acid." International Programme on Chemical Safety.
http://www.inchem.org/documents/pims/pharm/folicaci.htm (accessed on October 10, 2005).

"Folic Acid Fortification." U.S. Food and Drug Administration.
http://vm.cfsan.fda.gov/~dms/wh-folic.html (accessed on October 10, 2005).

"Vitamins: The Quest for Just the Right Amount." *Harvard Health Letter.* (June 2004): 1.

Formaldehyde

OTHER NAMES:
Methanal;
oxomethylene;
oxomethane;
methylene oxide;
formic aldehyde

FORMULA:
HCHO

ELEMENTS:
Carbon, hydrogen,
oxygen

COMPOUND TYPE:
Aldehyde (organic)

STATE:
Gas

MOLECULAR WEIGHT:
30.03 g/mol

MELTING POINT:
−92°C (−130°F)

BOILING POINT:
−19.1°C (−2.38°F)

SOLUBILITY:
Very soluble in water,
alcohol, ether, and
benzene

KEY FACTS

OVERVIEW

Formaldehyde (for-MAL-duh-hide) is a colorless, flammable gas with a strong, pungent odor that tends to polymerize readily. Polymerization occurs when individual molecules of formaldehyde combine with each other to make very large molecules called polymers. Over 4 billion kilograms (10 billion pounds) of formaldehyde were produced in the United States in 2004, the vast majority of which was used in the production of plastics and other polymers. To make handling and shipping easier and safer, the compound is usually provided as a 37 percent solution of formaldehyde in water to which has been added an additional 15 percent of methanol (methyl alcohol) to prevent polymerization.

Formaldehyde was first produced accidentally in 1859 by the Russian-French chemist Alexander Mikhailovich Butlerov (1828-1886). It was first synthesized in 1867 by the German chemist August Wilhelm von Hofmann (1818-1892) who was not, however, able to collect the compound in

CHEMICAL COMPOUNDS

Valley Christian
Junior High School
100 Skyway Dr.
San Jose, CA 95111

325

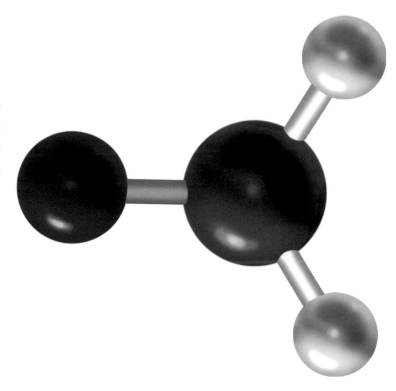

Formaldehyde. Red atom is oxygen; white atoms are hydrogen; and black atom is carbon. Gray sticks indicate double bonds. PUBLISHERS RESOURCE GROUP

pure form. That step was accomplished by German chemist Friedrich August Kekulé (1829-1896) in 1892.

HOW IT IS MADE

Formaldehyde occurs naturally in the atmosphere at a concentration of about 10 parts per billion (0.000 001%) partly as a by-product of plant and animal metabolism, and partly as a product of the reaction of sunlight with methane (CH_4), a much more abundant component of the air. At such low concentrations, it is not a natural source of the compound for commercial or industrial uses and is produced instead by the oxidation of methanol (methyl alcohol; CH_3OH) or gases extracted from petroleum (such as methane) over a catalyst of silver, copper, or iron with molybdenum oxide.

COMMON USES AND POTENTIAL HAZARDS

By far the most important application of formaldehyde is in the production of polymers and other organic chemicals.

Interesting Facts

- Formaldehyde was one of the first organic compounds to have been discovered in outer space.

- When some vegetables, such as cabbage and brussel sprouts are cooked, they emit small amounts of formaldehyde.

About one-quarter of commercial-use formaldehyde is used each year to make a family of polymers known as urea-formaldehyde resins, which are used to make dinnerware, particle board, fiber board, plywood, flexible foams, and insulation. Another 16 percent goes to the production of phenol-formaldehyde resins, with applications in molded and cast plastics, adhesives and bonding materials, laminating materials, brake linings, chemical equipment, machine housing, and a host of other applications. Smaller amounts of formaldehyde are used to make a variety of important chemicals including 1,4-butanediol, methylene diisocyanate, pentaerythritol, and hexamethylenetetramine. Other applications include use in controlled release fertilizers, in the production of nitroparaffin derivatives, in the treatment of textiles, and in the preservation of biological specimens. The last of these uses is probably well known to biology students; its use depends on the fact that formaldehyde kills most types of bacteria and can be used, therefore, to keep biological materials from decaying.

Formaldehyde poses a number of health hazards to humans and other animals. It may cause difficulty in breathing, headaches, fatigue, and lowered body temperature. At high levels of concentration or over long periods of exposure, formaldehyde can induce coma and death. Chronic exposure to formaldehyde is thought to be carcinogenic, producing tumors in the nose, throat, and respiratory system. People who work in factories where formaldehyde is used are at greatest risk for formaldehyde poisoning.

Formaldehyde is now known to be a potentially serious indoor air pollutant. So many products in a home contain

Words to Know

CARCINOGEN A chemical that causes cancer in humans or other animals.

CATALYST A material that increases the rate of a chemical reaction without undergoing any change in its own chemical structure.

METABOLISM A process that includes all of the chemical reactions that occur in cells by which fats, carbohydrates, and other compounds are broken down to produce energy and the compounds needed to build new cells and tissues.

POLYMER A compound consisting of very large molecules made of one or two small repeated units called monomers.

formaldehyde that significant levels of the compound may accumulate in a house. The primary sources of the formaldehyde are pressed wood products such as plywood and particleboard; furnishings; wallpaper; and durable press fabrics.

FOR FURTHER INFORMATION

"About Formaldehyde." Formaldehyde Council. http://www.formaldehyde.org/about_what.html (accessed on October 10, 2005).

"Formaldehyde (Methyl Aldehyde) Fact Sheet." Australian Government; Department of the Environment and Heritage. http://www.npi.gov.au/database/substance-info/profiles/45.html (accessed on October 10, 2005).

Gullickson, Richard. "Reference Data Sheet on Formaldehyde." Meridian Engineering & Technology. http://www.meridianeng.com/formalde.html (accessed on October 10, 2005).

"An Update on Formaldehyde-1997 Revision." U.S. Consumer Product Safety Commission. http://www.epa.gov/iaq/pubs/formald2.html (accessed on October 10, 2005).

See Also Urea

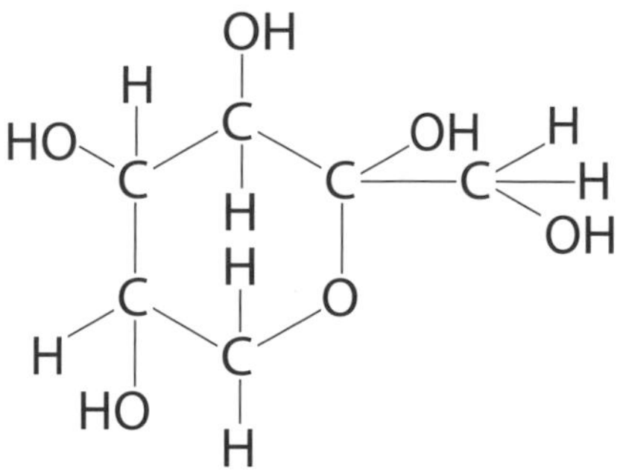

KEY FACTS

OTHER NAMES:
D-Fructose; fruit
sugar

FORMULA:
$C_6H_{12}O_6$

ELEMENTS:
Carbon, hydrogen,
oxygen

COMPOUND TYPE:
Carbohydrate
(organic)

STATE:
Solid

MOLECULAR WEIGHT:
180.16 g/mol

MELTING POINT:
103°C (217°F);
decomposes

BOILING POINT:
Not applicable

SOLUBILITY:
Very soluble in water
and acetone; soluble
in ethyl alcohol and
methyl alcohol

Fructose

OVERVIEW

Fructose (FROOK-tose) is a white crystalline solid found in honey and certain fruits and vegetables. It is the sweetest of the common sugars. Fructose is a carbohydrate, an organic compound in which five of the six carbon atoms are arranged in a ring to which are attached the hydrogen atoms and hydroxy (-OH) groups that make up the molecule. It is classified as a monosaccharide ("one sweet substance"), in contrast to sucrose, common table sugar, which is classified as a disaccharide ("two sweet substances"). Molecules of sucrose consist of two rings rather than the one ring found in fructose.

HOW IT IS MADE

Fructose is produced commercially by the hydrolysis of beet sugar or inulin, a polysaccharide found in the roots of a number of plants, including dahlias, Jerusalem artichokes, and chicory. Hydrolysis is the process by which a material is broken down into simpler elements by reacting it with

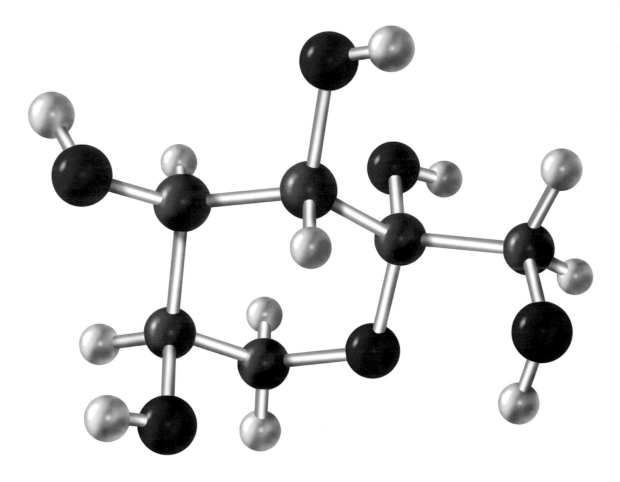

Fructose. Red atoms are oxygen; white atoms are hydrogen; and black atoms are carbon. PUBLISHERS RESOURCE GROUP

water. A polysaccharide is a carbohydrate with many simple sugar groups attached to each other. After hydrolysis of the beet sugar, or inulin, the resulting mixture is treated with lime (calcium oxide; CaO) to extract the fructose. It is then refined by removing impurities left from the preparation process.

COMMON USES AND POTENTIAL HAZARDS

Virtually the only important use of fructose is as a sweetener and preservative in a number of food products. In most cases, it is now used in the form of a substance known as high-fructose corn syrup (HFCS). HFCS was first introduced in the 1970s after scientists at the Clinton Corn Processing Company in Clinton, Iowa, developed a method of converting the sugar in corn into glucose and fructose. The Clinton

Interesting Facts

- Sucrose is a disaccharide that consists of one molecule of glucose joined to one molecule of fructose. The first step in the digestion of sucrose is the hydrolysis of the molecule, resulting in the formation of one molecule of glucose and one molecule of fructose. These compounds are the primary materials used by the human body in producing the energy needed to stay alive and grow.

process is relatively complicated. The polysaccharides in corn are first converted to glucose, and the glucose is then treated with enzymes that convert it to a thick syrup consisting of roughly half glucose and half fructose. From a nutritional standpoint, HFCS is very similar to sucrose, common table sugar, but it is less expensive to produce than sucrose and is more convenient to use in many instances.

High-fructose corn syrup has become one of the great success stories in the recent history of food processing in the United States. It has replaced sucrose in many applications, including nearly all soft drinks and fruit beverages, and in many jams and jellies, cookies, gum, baked goods, and other processed foods. Consumption of HFCS in the United States has increased from about 2 million metric tons (2.2 million short tons) in 1980 to just over 8 million metric tons (8.8 million short tons) in 2000.

When high-fructose corn syrup was first introduced, experts in nutrition did not anticipate that any health problems would be associated with the new product. After all, both fructose and glucose are naturally occurring substances that humans have been consuming for millennia. They are not so sure any more. Some evidence suggests that fructose is metabolized differently in the body than is glucose. It is not converted to energy as efficiently and may actually act more like a fat than like a sugar. Since few studies have been conducted on the ultimate chemical fat of fructose in the body, the nutritional value of HFCS is still the subject of some controversy among experts.

Words to Know

HYDROLYSIS The process by which a compound reacts with water to form two new compounds

The dangers of consuming HFCS are apparent for at least one group of people, those who lack the enzyme needed to metabolize fructose properly. Individuals with this genetic disorder may develop serious reactions if exposed to even a very small amount of fructose, reactions that include sweating, nausea, vomiting, confusion, abdominal pain, and, in extreme cases, convulsion and coma. Fortunately, this disorder is quite rare, affecting one person in about every twenty thousand individuals. For those with the disorder, however, care must be used in the kinds of sweeteners included in the diet.

FOR FURTHER INFORMATION

Basciano, Heather, Lisa Federico, and Khosrow Adeli. "Fructose, Insulin Resistance, and Metabolic Dyslipidemia." *Nutrition and Metabolism.* (Electronic journal) http://www.nutritionandmetabolism.com/content/2/1/5 (accessed on October 10, 2005).

Ophardt, Charles E. "Fructose." Elmhurst College. http://www.elmhurst.edu/~chm/vchembook/543fructose.html (accessed on October 10, 2005).

Squires, Sally. "Sweet but Not So Innocent?" *Washington Post.* (March 11, 2003): HE01. Available online at http://www.washingtonpost.com/ac2/wp-dyn/A8003-2003Mar10?language=printer (accessed on October 10, 2005).

"What Do We Know about Fructose and Obesity?" The International Food Information Council. http://www.ific.org/foodinsight/2004/ja/fructosefi404.cfm (accessed on October 10, 2005).

See Also Glucose; Sucrose

OTHER NAMES:
Benzene
hexachloride; BHC;
HCCH; HCH; TBH

FORMULA:
$C_6H_6Cl_6$

ELEMENTS:
Carbon, hydrogen,
chlorine

COMPOUND TYPE:
Chlorinated aromatic
hydrocarbon
(organic)

STATE:
Solid

MOLECULAR WEIGHT:
290.83 g/mol

MELTING POINT:
112.5°C (234.5°F)

BOILING POINT:
323.4°C (614.1°F)

SOLUBILITY:
Insoluble in water;
soluble in absolute
alcohol, chloroform,
and ether

KEY FACTS

Gamma-1,2,3,4,5, 6-Hexachlorocyclo- hexane

OVERVIEW

Gamma-1,2,3,4,5,6-hexachlorocyclohexane (GAM-uh one two three four five six HEK-sa-KLOR-oh-SYE-kloh-HEK-sane) exists in four isomeric forms. Isomers are two or more forms of a chemical compound with the same molecular formula, but different structural formulas and different chemical and physical properties. The isomer of greatest commercial interest is called the gamma (γ) isomer and is also known as lindane. It is sold commercially in more than a hundred commercial products under trade names such as 666, Africide, Agrocide, Aparasin, Arbitex, Ben-Hex, Bentox, Devoran, Entomoxan, Exagamma, Forlin, Gamaphex, Gammalin, Gammex, Gexane, Hexachloran, Isotox, Jacutin, Kwell, Lindafor, Lindatox, Lin-O-Sol, Lintox, Streunex, Tri-6, and Vitron.

Gamma-1,2,3,4,5,6-hexachlorocyclohexane is normally available as a white to yellowish powder or crystalline solid with a musty odor. Its color, odor, melting point, and other characteristics differ depending on the relative amount of the four isomers present in the final product. Gamma-1,2,3,4,5,

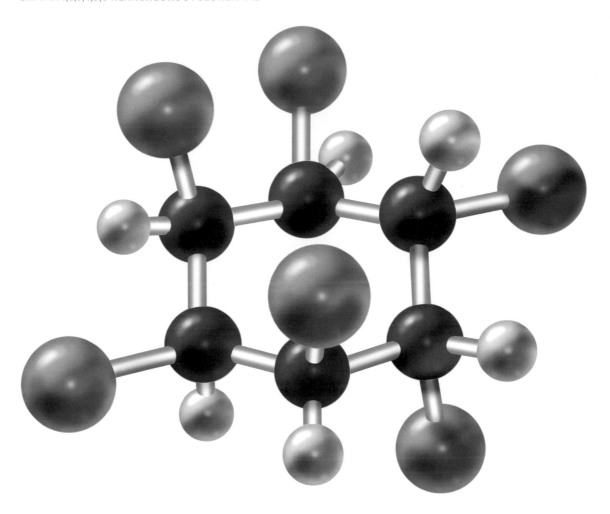

Gamma-1,2,3,4,5,6-Hexachlorocyclohexane. White atoms are hydrogen; black atoms are carbon; and green atoms are chlorine.
PUBLISHERS RESOURCE GROUP

6-hexachlorocyclohexane is non-flammable and stable in the presence of heat, light, strong acids, and carbon dioxide. The compound has two primary commercial uses: as a pesticide in agriculture and as a treatment for head lice, scabies, and other external parasites.

HOW IT IS MADE

Gamma-1,2,3,4,5,6-hexachlorocyclohexane is made by treating benzene (C_6H_6) with chlorine gas. In the process, each of the six hydrogen atoms in the benzene ring is replaced by a chlorine atom, resulting in the formation of a fully chlorinated benzene ring. That is, all of the original six hydrogen atoms have been replaced by chlorine atoms.

COMMON USES AND POTENTIAL HAZARDS

About 80 percent of all γ-1,2,3,4,5,6-hexachlorocyclohexane produced worldwide is used in agriculture, especially for treating soil and seeds. The wood and timber industries also use the product to protect trees from insects that attack them. In some places, γ-1,2,3,4,5,6-hexachlorocyclohexane is used as a spray to control the spread of mosquitoes. Veterinarians sometimes use the compound to treat or prevent fleas and other external parasites on animals. Gamma-1,2,3,4,5,6-hexachlorocyclohexane is also a major ingredient of products used to treat head lice, scabies, and similar pests that infest body hair.

Gamma-1,2,3,4,5,6-hexachlorocyclohexane is a member of the family of compounds known as the organochlorides, organic compounds that include one or more atom of chlorine in their molecular structure. The family also includes a number of other well-known products, such as DDT, dieldrin, aldrin, endrin, dinoseb, and chlordane. These compounds kill pests by attacking and incapacitating their nervous systems. Unfortunately, they have somewhat similar effects on the nervous systems of higher animals, including humans. Many of these organochlorides have been banned in commercial applications and many nations around the world.

The status of γ-1,2,3,4,5,6-hexachlorocyclohexane is still a matter of some controversy. It is banned for medicinal uses in many countries of the world, including Bangladesh, Belize, Bolivia, Brazil, Bulgaria, Chad, Denmark, Ecuador, Egypt, Finland, Guatemala, Honduras, Hong Kong, Hungary, Indonesia, Japan, Kuwait, Mozambique, New Zealand, The Netherlands, Nicaragua, Paraguay, the Republic of Korea, Singapore, Sweden, Taiwan, and Yemen. Its use is still permitted in the United States, although it has been banned as a known carcinogen in the state of California. The product has also been banned for some or all agriculture applications in more than 50 nations.

Gamma-1,2,3,4,5,6-hexachlorocyclohexane poses a health hazard to humans if ingested, inhaled, or deposited on the skin. It may cause skin irritation or rashes, nausea and vomiting, nervousness or irritability, accelerated heartbeat, convulsions or seizures, and dizziness or clumsiness. In rare cases, the ingestion of the compound has resulted in a person's death. It has also been classed as a likely carcinogen by the U.S. Environmental Protection Agency. The compound has been ranked in the top 10 percent among the most hazardous chemicals in ten out of eleven systems for making such rankings. It is ranked number thirty-two on the U.S. Agency for Toxic Substances and Disease Registry's list of 275 "priority" hazardous chemicals.

In spite of the apparent risk that γ-1,2,3,4,5,6-hexachlorocyclohexane poses to human health and the environment, its use is still permitted for many applications in the United States. A number of consumers' groups are currently working, however, to have the compound's use totally or partially banned.

FOR FURTHER INFORMATION

"Drug Information for Gamma Benzene Hexachloride." Drugs.com. http://www.drugs.com/cons/Gamma_benzene_hexachloride.html (accessed on October 10, 2005).

"Hexachlorocyclohexane (Mixed Isomers)." International Labour Organization. http://www.oit.org/public/english/protection/safework/cis/products/icsc/dtasht/_icsc04/icsc0487.htm (accessed on October 10, 2005).

"Lindane." IPCS International Programme on Chemical Safety. http://www.inchem.org/documents/hsg/hsg/hsg054.htm (accessed on October 10, 2005).

"Lindane Education and Research Network." National Pediculosis Association. http://www.headlice.org/lindane/ (accessed on October 10, 2005).

OTHER NAMES:
Gelatine

FORMULA:
Not applicable

ELEMENTS:
Carbon, hydrogen, oxygen, nitrogen, and others

COMPOUND TYPE:
Not applicable

STATE:
Solid

MOLECULAR WEIGHT:
Not applicable

MELTING POINT:
Not applicable

BOILING POINT:
Not applicable

SOLUBILITY:
Soluble in hot water and glycerol; insoluble in most organic solvents

Gelatin

OVERVIEW

Gelatin (JELL-ah-tin) is a mixture, not a compound. Mixtures differ from compounds in a number of important ways. The parts that make up a mixture are not chemically combined with each other, as they are in a compound. Also, mixtures have no definite composition, but consist of varying amounts of the substances from which they are formed. Gelatin is a mixture of water-soluble proteins with high molecular weights. It typically occurs as a brittle solid in the form of colorless or slightly yellow flakes or sheets, or in powder form, with virtually no taste or odor. It absorbs up to ten times its own weight when mixed with cold water and dissolves in hot water. When a solution of gelatin in hot water is cooled, it takes the form of a gel, a jelly-like material perhaps most commonly seen as the popular dessert called JELL-O™. Gelatin is also available in a number of other commercial forms, such as Knox Gelatin™, Puragel®, and Gelfoam®. Gelatin has been known to humans for many centuries, but it was not widely marketed until the late

1890s. Its name comes from the Latin word *gelatus*, which means "frozen."

HOW IT IS MADE

Gelatin is made by boiling animal parts with high protein content, such as skin, ligaments, tendons, cartilage, and bones. The boiling process breaks down molecular bonds between individual collagen strands in the animal tissue. Collagen is a structural protein found in bone, cartilage, and connective tissue. The collagen formed by this process can be further disintegrated through additional boiling with either acid or alkali. Type A gelatin is produced when collagen is boiled in an acidic solution, and type B gelatin is produced by boiling collagen in an alkaline solution.

Most of the animal parts used to make gelatin come from cattle and pigs and are left over from meat and leather processing. Gelatin can also be made from fish. One of the oldest forms of gelatin is isinglass, made from the swim bladders of fish. Jewish and Muslim dietary laws prohibit believers from eating pork, so some gelatin is made without pig parts. Vegetarians and vegans do not eat any animal products, so gelatin manufacturers also make similar products using vegetable carbohydrates, such as agar and pectin. These vegetarian gelatins are not true gelatin, which is always made from animal proteins.

COMMON USES AND POTENTIAL HAZARDS

People discovered gelatin centuries ago and experimented with various uses for it. In the early 1800s, for example, gelatin was included in the food served to French soldiers as a source of dietary proteins. In the 1890s, Knox Gelatin™ was sold as a cure for dry fingernails. Manufacturers claimed that dry fingernails were caused by a lack of protein and that eating gelatin would cure the condition. No scientific evidence exists for that claim, but Knox Gelatin™ became popular among consumers nonetheless.

In 1900, the Genesee Pure Food Company began selling flavored gelatin under the name JELL-O™. In the early 1900s, the company began distributing booklets containing recipes using JELL-O™, eventually giving out more than 15 million such booklets. JELL-O™ eventually became one of the most

Interesting Facts

- Before refrigerators became common, gelatin was used to keep foods fresh and attractive. Packing a food in gelatin prevents oxygen from reacting with the food and causing spoilage.

- Synchronized swimmers sometimes use gelatin to hold their hair in place during performances.

- Aspic is a clear jelly often made with gelatin. It is a component of many elegant dishes, one of which, "Oeufs de caile en aspic et caviar" (Quail eggs in aspic with caviar), was served in first class on the doomed steam ship *Titanic* in 1912.

popular desserts in the United States and other countries. It has been used to make a variety of pleasant tasting, attractive looking desserts molded into many different shapes. Cooks have combined gelatin with water, milk, soft drinks, other liquids, whipped toppings, or mayonnaise to change its taste and texture. The product is often served with fruits or vegetables as a salad. Gelatin is also combined with marshmallows, jelly-beans, jelly, yogurt, gummy candies, ice cream, and margarine to produce desserts of many textures and flavors. The product is sometimes recommended as a fat substitute because it provides volume in a diet without adding many calories. Some people include gelatin products in their diets as a way of increasing protein intake. Although plain gelatin is almost entirely protein, it actually has relatively little nutritional value.

Gelatin has many other uses, including:

- As a raw material for the manufacture of capsules and gels in the production of drugs;

- As a way of holding silver halide (silver bromide and silver iodide) crystals in place on photographic films and plates;

- In the manufacture of blocks used to determine the possible effects of various types of ammunition on human flesh;

Words to Know

ALKALI A strong base.

MIXTURE A collection of two or more elements and/or compounds with no definite composition.

PROTEIN A large, complex compound made of long chains of amino acids. Proteins have a number of essential functions in living organisms.

- As a binder that holds sand on sandpaper or to make certain types of paper products (such as playing cards) bright and shiny;

- As an additive in various types of cosmetics and skin treatments;

- In the manufacture of meshes used in the repair of wounds and in the production of artificial heart valves;

- In the production of certain types of cement;

- For the manufacture of light filters used in theatrical productions and for other specialized purposes;

- As a culturing medium for bacteria;

- As a stabilizer and thickener for certain types of foods, especially ice cream and some other dairy products;

- In the manufacture of printing inks;

- As an additive in the production of plastics and rubber products.

FOR FURTHER INFORMATION

"Gelatin." WholeHealthMC.com. http://www.wholehealthmd.com/refshelf/substances_view/ 1,1525,10151,00.html (accessed on December 22, 2005).

"A History of JELL-O™ Brand." Kraftfoods.com. http://www.kraftfoods.com/jello/main.aspx?s=&m=jlo_history (accessed January 4, 2006).

"The Jell-O Museum." The Genesee Pure Food Co. http://www.jellomuseum.com/#Page1. (accessed on December 22, 2005).

"What Exactly Is Jell-O Made From?" How Stuff Works.
   http://home.howstuffworks.com/question557.htm   (accessed
   on December 22, 2005).

"What Is Gelatin." PB Leiner.
   http://www.gelatin.com/ (accessed on December 22, 2005).

Wyman, Carolyn. *JELL-O: A Biography.* Fort Washington, PA:
   Harvest Books, 2001.

See Also Collagen

Glucose

OVERVIEW

Glucose (GLOO-kose) is a simple sugar used by plants and animals to obtain the energy they need to stay alive and to grow. It is classified chemically as a monosaccharide, a compound whose molecules consist of five- or six-membered carbon rings with a sweet flavor. Other common examples of monosaccharides are fructose and galactose. Glucose usually occurs as a colorless to white powder or crystalline substance with a sweet flavor. It consists in two isomeric forms known as the D configuration and the L configuration. Dextrose is the common name given to the D conformation of glucose.

Credit for the discovery of glucose is often given to the German chemist Andreas Sigismund Marggraf (1709-1782). In 1747, Marggraf isolated a sweet substance from raisins that he referred to as *einer Art Zücker* (a kind of sugar) that we now recognize as glucose. More than 60 years later, the German chemist Gottlieb Sigismund Constantine Kirchhof (1764-1833) showed that glucose could also be obtained from

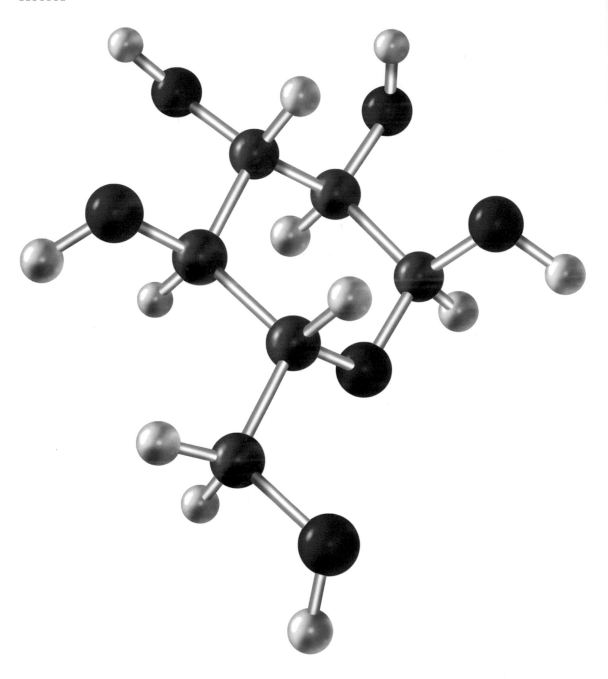

Glucose. Red atoms are oxygen; whjte atoms are hydrogen; and black atoms are carbon. PUBLISHERS RESOURCE GROUP

the hydrolysis of starch and that starch itself was nothing other than a very large molecule (polysaccharide) composed of many repeating glucose units. The molecular structure for glucose was finally determined in the 1880s by German

Interesting Facts

- The name glucose comes from the Greek word *gleucos* for "sweet wine."

chemist Emil Fischer (1852-1919), part of the reason for which he was awarded the 1902 Nobel Prize in chemistry.

HOW IT IS MADE

Glucose is synthesized naturally in plants and some single-celled organisms through the process known as photosynthesis. In this process, sunlight catalyzes the reaction between carbon dioxide and water that results in the formation of a simple carbohydrate (glucose) and oxygen. The overall reaction can be summarized by a rather simple chemical equation:

$$6CO_2 + 6H_2O \rightarrow C_6H_{12}O_6 + 6O_2$$

However, photosynthesis actually involves a number of complex reactions that occur in two general phases, the light reactions and the dark reactions.

Glucose is produced commercially through the steam hydrolysis of cornstarch or waste products containing cellulose (a large molecule composed of glucose units) using a dilute acid catalyst. The product thus obtained is typically not very pure, but is contaminated with maltose (a disaccharide consisting of two molecules of glucose joined to each other) and dextrins (larger molecules consisting of a number of glucose units joined to each other).

COMMON USES AND POTENTIAL HAZARDS

Glucose is the primary chemical from which plants and animals derive energy. In cells, glucose is broken down in a complex series of reactions to produce energy with carbon dioxide and water as byproducts.

Words to Know

CATALYST A material that increases the rate of a chemical reaction without undergoing any change in its own chemical structure.

HYDROLYSIS The process by which a compound reacts with water to form two new compounds.

ISOMERS Two or more forms of a chemical compound with the same molecular formula, but different structural formulas and different chemical and physical properties.

METABOLISM A process that includes all of the chemical reactions that occur in cells by which fats, carbohydrates, and other compounds are broken down to produce energy and the compounds needed to build new cells and tissues.

Glucose also has a number of commercial uses, nearly all of them related to the food processing business. It is used in the production of confectionary products; chewing gum; soft drinks; ice creams; jams, jellies, and fruit preparations; baby foods; baked products; and beers and ciders. A relatively small amount is used for non-food purposes, primarily in the production of other organic chemicals, such as citric acid, the amino acid lysine, insulin, and a variety of antibiotics.

The most important health problem associated with glucose is diabetes. Diabetes is a medical condition that develops when the body either does not produce adequate amounts of insulin or cannot use that compound properly. Insulin is a hormone that controls the metabolism of glucose in the body. If glucose is not metabolized properly, a person's body acts as if it is "starving." Symptoms of diabetes include excessive hunger, weight loss, and exhaustion. If left untreated, the condition can result in coma and death. Diabetics must have an artificial source of insulin (usually from injections) and watch their diets to keep these symptoms under control.

FOR FURTHER INFORMATION

"All about Diabetes." American Diabetes Association. http://www.diabetes.org/about-diabetes.jsp (accessed on October 10, 2005).

"Carbohydrates." Kimball's Biology Pages.
http://users.rcn.com/jkimball.ma.ultranet/BiologyPages/C/
Carbohydrates.html (accessed on October 10, 2005).

"Dextrose, Anhydrous." J. T. Baker.
http://www.jtbaker.com/msds/englishhtml/D0835.htm
(accessed on October 10, 2005).

"Glucose." Department of Chemistry, Imperial College London.
http://www.ch.ic.ac.uk/vchemlib/mim/bristol/glucose/
glucose_text.htm (accessed on October 10, 2005).

See Also Fructose; Sucrose

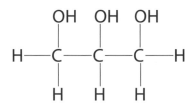

Glycerol

OVERVIEW

Glycerol (GLIH-ser-ol) is a clear, colorless, odorless, sweet-tasting syrupy liquid. It is a trihydric alcohol, meaning that its molecules contain three hydroxyl (-OH) groups. Glycerol occurs naturally in all animal and plant cells. Glycerol was discovered in 1779 by the German chemist Karl Wilhelm Scheele (1742-1786) and named by the French chemist Michel Eugéne Chevreul (1786-1889) because of its sweet taste (*glycos* means "sweet" in Greek). In 1836 the French chemist Théophile-Jules Pelouze (1807-1867) determined the molecular formula for glycerol, and three decades later, in 1872, the compound was first synthesized (created in a laboratory) by the French chemist Charles Friedel (1832-1899). About 250 million kilograms (500 million pounds) of glycerol are produced in the United States each year, the majority of which goes to the production of food and personal care products.

HOW IT IS MADE

The traditional method of making glycerol is by the saponification of fats. Fats are the esters of glycerol and

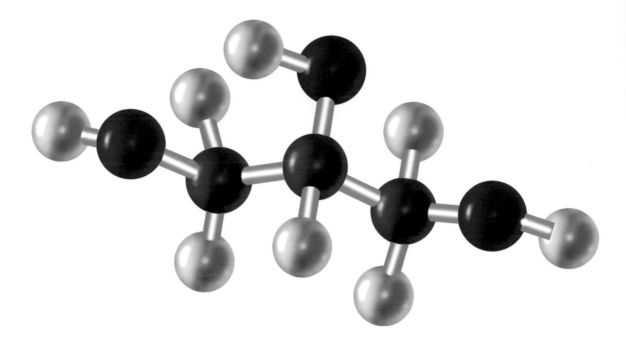

one or more alcohols. When a fat is hydrolyzed in the presence of a catalyst, it is converted into the glycerol and alcohols from which it was originally made. This type of hydrolysis is called saponification because it is the usual method by which soaps (the Latin word for "soap" is *sapon*). Glycerol can be obtained, then, as a byproduct in the manufacture of soaps.

A number of synthetic methods for making glycerol are also available. Most of these procedures begin with propylene (propene; $CH_2=CHCH_3$) and chlorine and involve a series of steps that convert the three-carbon propylene to the three-carbon glycerol. Glycerol can also be obtained by treating simple sugars with hydrogen over a nickel catalyst. In the United States, about 80 percent of all glycerol produced is obtained by saponification methods, and the remaining 20 percent is made by synthetic methods.

COMMON USES AND POTENTIAL HAZARDS

The most common use for glycerol in the United States is in food products, where it acts as a sweetener and as a thickener in many foods. For example, it is added to ice cream to improve texture and to candy products and baked

Interesting Facts

- As winter approaches, many insects begin to produce glycerol to replace the water in their body tissues. Glycerol acts as an antifreeze to prevent the insects from freezing during the coldest part of the year.

goods to increase the sweetness of the product. It is also used to make the flexible coatings on cheeses, sausages, and other meat products. About one-quarter of all glycerol made in the United States is used in the food products industry.

Nearly the same amount of glycerol is used in the preparation of personal care products, such as skin, hair, and soap products (23 percent) and in oral hygiene products, such as toothpastes and mouthwashes (17 percent). Some of the products that include glycerol are moisturizers, detergents, soaps, hair coloring agents, mascara, nail polish, nail polish removers, perfumes, body lotions, hair sprays, shaving creams, lipsticks, cough medicines, shampoos, and hair conditioners. Glycerol is also used as a humectant in tobaccos. A humectant is a material that helps a product conserve moisture and prevent it from drying out.

Other uses for glycerol include:

- In the manufacturer of explosives;

- In the production of a variety of plastics and polymers, such as polyether polyols, urethanes, and alkyd resins;

- As a lubricant in pumps, bearings, gaskets, and other mechanical systems;

- In the manufacture of ink rolls, inks, and rubber stamps;

- As an emulsifying agent, a material that helps two liquids that are not soluble in each other to stay mixed;

- As an antifreeze; and

- In a number of medical applications, such as the treatment of glaucoma and stroke.

Words to Know

ESTER An organic compound formed in the reaction between an organic acid and an alcohol.

HYDROLYSIS The process by which a compound reacts with water to form two new compounds.

MISCIBLE able to be mixed; especially applies to the mixing of one liquid with another.

SYNTHESIS A chemical reaction in which some desired chemical product is made from simple beginning chemicals, or reactants.

Glycerol poses some safety problems because it is combustible and explosive under certain conditions. It presents no health hazards under normal circumstances of use, however.

FOR FURTHER INFORMATION

"Glycerol." International Chemical Safety Cards. http://www.healthy-communications.com/msdsglycerin1.html (accessed on October 10, 2005).

"Glycerol." PDRhealth. http://www.pdrhealth.com/drug_info/nmdrugprofiles/ nutsupdrugs/gly_0304.shtml (accessed on October 10, 2005).

Legwold, Gary. "Hydration Breakthrough." *Bicycling* (July 1994): 72-73.

"Material Safety Data Sheet: Glycerine." Department of Chemistry, Iowa State University. http://avogadro.chem.iastate.edu/MSDS/glycerine.htm (accessed on October 10, 2005).

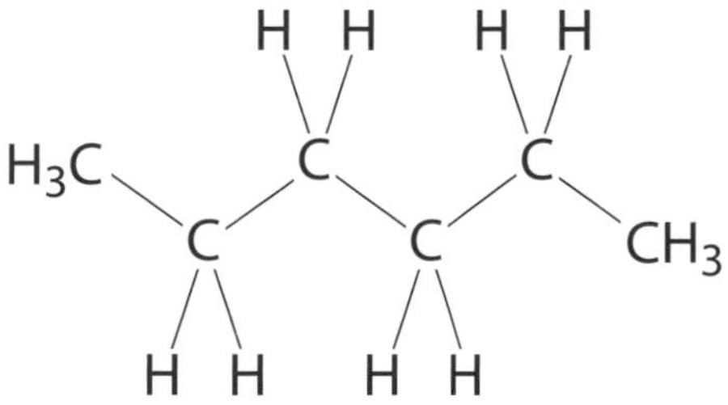

OTHER NAMES:
n-hexane

FORMULA:
C_6H_{14}

ELEMENTS:
Carbon, hydrogen

COMPOUND TYPE:
Alkane; saturated
hydrocarbon
(organic)

STATE:
Liquid

MOLECULAR WEIGHT:
86.18 g/mol

MELTING POINT:
−95.35°C (−139.6°F)

BOILING POINT:
68.73°C (155.7°F)

SOLUBILITY:
Insoluble in water;
very soluble in ethyl
alcohol; soluble in
ether and chloroform

Hexane

OVERVIEW

Hexane (HEX-ane) is a colorless flammable liquid with a faint petroleum-like odor. Chemically it is classified as a saturated hydrocarbon, which means that its molecules contain only carbon and hydrogen atoms joined only by single bonds. Saturated hydrocarbons are also known as alkanes. By far its most important use is as a solvent in a variety of industrial operations.

HOW IT IS MADE

Hexane is extracted from petroleum. Petroleum is a complex mixture of solid, liquid, and gaseous hydrocarbons that has virtually no use itself. However, the fractional distillation of petroleum produces hundreds of individual compounds, each of which has its own important commercial and industrial applications. Fractional distillation is the process by which petroleum is heated in tall towers. The components of petroleum boil off at different temperatures, rise to

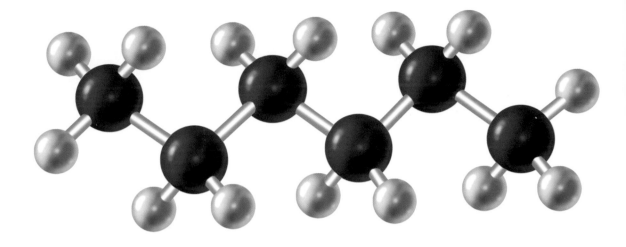

Hexane. White atoms are hydrogen and black atoms are carbon. PUBLISHERS RESOURCE GROUP

different heights in the tower, and are condensed at different levels above the base of the tower. As a liquid with a low boiling point, hexane boils off and rises to upper levels of the tower, where it condenses and is removed in a portion of the petroleum known as petroleum ether or ligroin. Hexane can then be separated from other constituents of petroleum ether by a second distillation, in which each component boils off and is condensed at its own distinctive boiling point.

COMMON USES AND POTENTIAL HAZARDS

By far the most important use of hexane is in solvents used for a variety of purposes, such as the extraction of oils from seeds and vegetables; as a degreaser and cleaning agent for printing equipment; as a solvent in glues such as rubber cement; as an ingredient in inks and varnishes; in the shoe and leather manufacturing industry; and in the roofing industry.

Hexane poses both safety and health risks for humans and other animals. The liquid vaporizes easily and the vapors formed ignite easily and may even explode under the proper conditions. The primary health hazard related to hexane occurs by breathing in the compound. When inhaled, it can cause numbness in the hands and feet, weakness in the feet and lower legs, paralysis of the arms and legs, muscle wasting,

Interesting Facts

- Hexane to which a red or blue dye has been added is sometimes used to make thermometers used for measuring low temperatures.

damage to nerves, nausea and vomiting, jaundice, skin rashes, irritation of the eyes and throat, blurred vision, mental confusion, and coma. There is no evidence, however, that hexane is carcinogenic.

The health hazards posed by hexane are of concern in only two circumstances: among workers who handle the liquid on a regular basis; and among people who deliberately inhale the compound as part of a solvent for the purpose of getting "high." In both cases, a person is exposed to much higher concentrations of hexane that one would encounter in commercial or industrial products. The practice of glue-sniffing, which has become popular among some teenagers, can result in serious health problems, including dizziness, increased heart rate, high blood pressure, disorientation, confusion, and occasionally violent impulses or suicide attempts.

Words to Know

CARCINOGEN A substance that causes cancer in humans or other animals.

DISTILLATION A process of separating two or more substances by boiling the mixture of which they are composed and conden-sing the vapors produced at different temperatures.

SOLVENT A liquid in which another substance is dissolved, to form a solution.

FOR FURTHER INFORMATION

"Hexane." U.S. Environmental Protection Agency Technology Transfer Network, Air Toxics Website. http://www.epa.gov/ttn/atw/hlthef/hexane.html (accessed on October 12, 2005).

Menhard, Francha Roffe. The Facts about Inhalants. New York: Benchmark Books, 2004.

"ToxFAQs for n-hexane." Agency for Toxic Substances and Disease Registry. http://www.atsdr.cdc.gov/tfacts113.html (accessed on October 12, 2005).

H
|
Cl

OTHER NAMES:
Anhydrous
hydrochloric acid

FORMULA:
HCl

ELEMENTS:
Hydrogen; chlorine

COMPOUND TYPE:
Inorganic acid

STATE:
Gas

MOLECULAR WEIGHT:
36.46 g/mol

MELTING POINT:
−114.17°C (−173.51°F)

BOILING POINT:
−85°C (−121°F)

SOLUBILITY:
Very soluble in
water; soluble in
alcohol and ether

KEY FACTS

Hydrogen Chloride

OVERVIEW

Hydrogen chloride (HY-druh-jin KLOR-ide) is a colorless gas with a strong, suffocating odor. The gas is not flammable, but is corrosive, that is, capable of attacking and reacting with a large variety of other compounds and elements. Hydrogen chloride is most commonly available as an aqueous solution known as hydrochloric acid. It is one of the most important industrial chemicals in the world. In 2004, just over 5 million metric tons (5.5 million short tons) of hydrogen chloride were produced in the United States, making it the eighteenth most important chemical in the nation for that year.

Hydrogen chloride has probably been known as far back as the eighth century, when the Arabian chemist Jabir ibn Hayyan (c. 721-c. 815; also known by his Latinized name of Geber) described the production of a gas from common table salt (sodium chloride; NaCl) and sulfuric acid (H_2SO_4). The compound was mentioned in the writings of a number of alchemists during the Middle Ages and was probably first

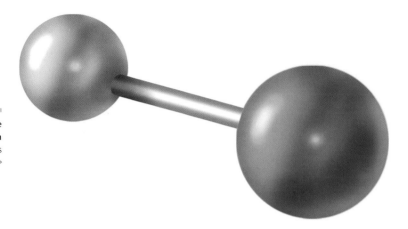

Hydrogen chloride. White atom is hydrogen and green atom is chlorine. PUBLISHERS RESOURCE GROUP

produced in a reasonably pure form by the German chemist Johann Rudolf Glauber (1604-1670) in about 1625. The first modern chemist to prepare hydrogen chloride and describe its properties was the English chemist Joseph Priestley (1733-1804) in 1772. Forty years later, in 1818, the English chemistry and physicist Humphry Davy (1778-1829) showed that the compound consisted of hydrogen and chlorine, giving it the correct formula of HCl.

Commercial production of hydrogen chloride had its beginning in Great Britain in 1823. The method of production most popular there and, later, throughout Europe was one originally developed by the French chemist Nicholas Leblanc (1742-1806) in 1783. Leblanc had invented the process as a method for producing sodium hydroxide and sodium carbonate, two very important industrial chemicals. Hydrogen chloride was produced as a byproduct of the Leblanc process, a byproduct for which there was at first no use. The gas was simply allowed to escape into the air. The suffocating and hazardous release of hydrogen chloride prompted governments to pass legislation requiring some other means of disposal for the gas. In England, that law was called the Alkali Act and was adopted by the parliament in 1863. Unable to release hydrogen chloride into the air, manufacturers began dissolving it in water and producing hydrochloric acid. Before long, a number of important commercial and industrial uses for the acid itself were discovered. The "useless" byproduct of the Leblanc process soon became as important as the primary products of the process, sodium hydroxide and sodium carbonate.

Interesting Facts

- Hydrogen chloride was studied by the famous alchemist Basil Valentine (c. 1394–?), who gave it the name *spiritus salis* ("the spirit of salt") by which it was known to most alchemists.

- Hydrochloric acid has traditionally been known as muriatic acid, a name that is still sometimes used by workers in fields in which it is used.

HOW IT IS MADE

Hydrogen chloride is still sometimes made today by the traditional process of reacting sodium chloride (NaCl) with a sulfate, such as sulfuric acid or iron(II) sulfate ($FeSO_4$). However, more than 90 percent of the hydrogen chloride produced throughout the world today comes as the byproduct of the chlorination of organic compounds. Chlorination is the process by which chlorine gas reacts with an organic compound, usually replacing some of the hydrogen present in the compound. Since a large number of important chlorinated organic compounds are produced each year, large amounts of hydrogen chloride gas are produced as a byproduct. That gas is simply removed from the reaction and stored in cylinders for future use. Other methods of producing hydrogen chloride include the direct synthesis of hydrogen gas and chlorine gas (producing a very pure product) and the reaction of sodium chloride, sulfur dioxide, oxygen, and water with each other at high temperatures (the Hargreaves process).

COMMON USES AND POTENTIAL HAZARDS

Hydrogen chloride and hydrochloric acid have some uses in common, and some that are different from each other. In both dry and liquid form, the largest single use of hydrogen chloride is in the synthesis of organic and inorganic chlorides. A large number of compounds important in commerce and industry contain chlorine, including most pesticides, many pharmaceuticals, and a number of polymeric products.

Words to Know

ALCHEMY An ancient field of study from which the modern science of chemistry evolved.

ANHYDROUS Without water or moisture.

AQUEOUS A solution is one that consists of some material dissolved in water.

CATALYST A material that increases the rate of a chemical reaction without undergoing any change in its own chemical structure.

POLYMER A compound consisting of very large molecules made of one or two small repeated units called monomers.

Hydrochloric acid is also used widely in the processing of metallic ores and the pickling of metals. Pickling is the process by which a metal is cleaned, usually with an acid, to remove rust and other impurities that have collected on the metal. Some additional uses of hydrogen chloride and hydrochloric acid include the following:

- In the brining of foods and other materials. Brining is the process by which a material is soaked in a salt solution, usually in order to preserve the material;

- In the treatment of swimming pool water;

- As a catalyst in industrial chemical reactions;

- In the manufacture of semiconductors and other electronic components;

- To maintain the proper acidity in oil wells (to keep oil flowing smoothly);

- For the etching of concrete surfaces;

- In the production of aluminum, titanium, and a number of other important metals.

Both hydrogen chloride and hydrochloric acid pose serious health risks to humans and other animals. The gas is an irritant to the eyes and respiratory system, causing coughing, choking, and tearing, as well as more serious damage to tissues. Hydrochloric acid can burn the skin and mucous membranes. Exposure of only five parts per million of the gas can produce noticeable symptoms of distress, and exposure of more than 2,000 parts per million can be fatal. If hydrochloric acid gets into the eyes, blindness may result. Since hydrochloric acid is present in many household products, users should exercise great care when working with such materials.

FOR FURTHER INFORMATION

"Hydrochloric Acid." National Safety Council.
   http://www.nsc.org/library/chemical/Hydrochl.htm (accessed on October 12, 2005).

"Hydrogen Chloride." Agency for Toxic Substances and Disease Registry.
   http://www.atsdr.cdc.gov/tfacts173.pdf (accessed on October 12, 2005).

"Hydrogen Chloride."
   http://www.ucc.ie/ucc/depts/chem/dolchem/html/comp/hcl.html (accessed on October 12, 2005).

"Hydrogen Chloride, HCl." Defense Service Center.
   http://www.c-f-c.com/gaslink/pure/hydrogen-chloride.htm (accessed on October 12, 2005).

See Also Sodium Chloride

H–O–O–H

OTHER NAMES:
Hydrogen dioxide;
hydroperoxide;
peroxide

FORMULA:
H_2O_2

ELEMENTS:
Hydrogen; oxygen

COMPOUND TYPE:
Oxide (inorganic)

STATE:
Liquid

MOLECULAR WEIGHT:
34.02 g/mol

MELTING POINT:
$-0.43°C$ $(-31°F)$

BOILING POINT:
150.2°C (302.4°F)

SOLUBILITY:
Very soluble in water;
soluble in ether

Hydrogen Peroxide

OVERVIEW

Hydrogen peroxide (HY-druh-jin per-OK-side) is a clear, colorless, somewhat unstable liquid with a bitter taste. When absolutely pure, the compound is quite stable. Even small amounts of impurities (such as iron or copper), however, act as catalysts that increase its tendency to decompose, sometimes violently, into water and nascent oxygen (O). To prevent decomposition, small amounts of inhibitors, such as acetanilide or sodium stannate are added to pure hydrogen peroxide and hydrogen peroxide solutions.

Hydrogen peroxide was discovered in 1818 by French chemist Louis Jacques Thénard (1777-1857). It was first used commercially in the 1800s, primarily to bleach hats. Today, industrial processes make about 500 million kilograms (1 billion pounds) of hydrogen peroxide annually for use in a wide variety of applications ranging from whitening of teeth to propelling rockets.

Hydrogen peroxide. Red atoms are oxygen and white atoms are hydrogen. PUBLISHERS RESOURCE GROUP

HOW IT IS MADE

Hydrogen peroxide occurs in very small amounts in nature. It is formed when atmospheric oxygen reacts with water to form H_2O_2. Hydrogen peroxide is also present in plant and animal cells as the byproduct of metabolic reactions that occur in those cells.

The large amounts of hydrogen peroxide used in industry are prepared in a complex series of reactions that begins with any one of a family of compounds known as the alkyl anthrahydroquinones, such as ethyl anthrahydroquinone. The anthrahydroquinones are three-ring compounds that can be converted back and forth between two or more similar structures. During the conversion from one structure to another, hydrogen peroxide is produced as a byproduct. The anthraquinone is continuously regenerated during the production of hydrogen peroxide, making the process very efficient.

Other methods for the preparation of hydrogen peroxide are also available. For example, the electrolysis of sulfuric acid results in the formation of a related compound, peroxysulfuric acid (H_2SO_5), which then reacts with water to form hydrogen peroxide. A third method of preparation involves the heating of isopropyl alcohol [2-propanol; $(CH_3)_2CHOH$] at high temperature and pressure, resulting in the formation of hydrogen peroxide as one product of the reaction.

Interesting Facts

- Hydrogen peroxide is sold in concentrations ranging from 3 percent (for home use) to 90 percent (for industrial applications).

- Scientists have discovered hydrogen peroxide in the atmosphere of Mars.

COMMON USES AND POTENTIAL HAZARDS

Most of hydrogen peroxide's applications depend on the fact that it tends to break down, releasing a single atom of nascent oxygen (O):

$$H_2O_2 \rightarrow H_2O + (O)$$

The term nascent oxygen refers to a single atom of oxygen, a structure that is chemically very active. Nascent oxygen tends to be a very strong oxidizing agent. For example, the use of hydrogen peroxide with which most people are probably familiar is as an antiseptic, a substance used to kill germs. Hydrogen peroxide achieves this result because the nascent oxygen it releases destroys bacteria, fungi, and other microorganisms that cause disease.

The most important industrial application of hydrogen peroxide—its use in the pulp and paper industry—also depends on its oxidizing properties. In this case, it is used to bleach the materials of which paper is made, converting colored compounds to colorless compounds. About 55 percent of all hydrogen peroxide made in the United States is used for this purpose. Another nine percent is used in the bleaching of other materials, such as textiles, furs, feathers, and hair. Another important application of hydrogen peroxide is in water and sewage treatment plants, where its antibacterial action destroys disease-causing organisms in the water. Some additional uses of hydrogen peroxide include:

- In bakeries to condition dough and make it easier to work with;

- For cleaning metals;

- As a rocket propellant;

Words to Know

CATALYST A material that increases the rate of a chemical reaction without undergoing any change in its own chemical structure.

INHIBITOR A substance added to another substance to prevent or slow down an unwanted reaction.

METABOLISM A process that includes all of the chemical reactions that occur in cells by which fats, carbohydrates, and other compounds are broken down to produce energy and the compounds needed to build new cells and tissues.

OXIDATION A chemical reaction in which oxygen reacts with some other substance or, alternatively, in which some substances loses electrons to another substance, the oxidizing agent.

- In the preparation of other organic and inorganic compounds;

- As a neutralizing agent in the production of wines; and

- As a disinfectant in the treatment of seeds for agricultural purposes.

The hydrogen peroxide solutions with which people come into contact at home pose little or no health hazard because the concentration of the compound is very low, usually about 3 percent. Prolonged use of hydrogen peroxide may cause burns on the skin, however, and the more concentrated solutions used in industry present more serious hazards. They can be toxic if ingested and are explosive if not stored properly.

FOR FURTHER INFORMATION

"Hydrogen Peroxide (>60% Solution in Water)." International Labour Organization.
http://www.ilo.org/public/english/protection/safework/cis/products/icsc/dtasht/_icsc01/icsc0164.htm
(accessed on October 12, 2005).

"Introduction to Hydrogen Peroxide." U.S. Peroxide.
http://www.h2o2.com/intro/overview.html
(accessed on October 12, 2005).

"A Prescription for Death?" CBSNews.com
http://www.cbsnews.com/stories/2005/01/12/60II/main666489.shtml (accessed on October 12, 2005).

$$Fe = O$$

KEY FACTS

Iron(II) Oxide

OVERVIEW

Iron(II) oxide (EYE-urn two OK-side) is a black, dense powder that occurs in nature as the mineral wustite. It reacts readily with oxygen in the air to form iron(III) oxide and with carbon dioxide to form iron(II) carbonate ($FeCO_3$).

HOW IT IS MADE

Iron(II) oxide occurs in nature as the result of the incomplete oxidation of iron metal. It can be prepared synthetically by heating iron(II) oxalate (FeC_2O_4), although the product of this reaction is contaminated with another oxide of iron, triiron tetroxide (Fe_3O_4).

COMMON USES AND POTENTIAL HAZARDS

Iron(II) oxide has three primary uses. First, it has long been used as a dye or pigment in pottery, glazes, and glasses, especially in the green, heat-absorbing glass used in buildings, automobiles, and other applications. The compound is

Iron(II) oxide. Red atom is oxygen and orange atom is iron. Gray stick is a double bond. PUBLISHERS RESOURCE GROUP

also used as a raw material in the production of steel. Finally, iron(II) oxide is used as a catalyst in a number of industrial and chemical operations.

Inhalation of iron(II) oxide fumes or dust is considered a hazard and can cause throat and nasal irritation. High levels of exposure may lead to a condition known as metal fume fever, a workplace exposure illness that causes flu-like symptoms. Continued exposure at high levels can have more serious effects, including a disease known as siderosis, an inflammation of the lungs that is accompanied by pneumonia-like symptoms.

Interesting Facts

- Both iron(II) oxide and its cousin, iron(III) oxide are abundant in rocks on the surface of Mars. The former accounts for the dark, nearly jet black, color of some rocks, while the latter is responsible for the predominant red color of the planet's surface.

Words to Know

OXIDATION A chemical reaction in which oxygen reacts with some other substance or, alternatively, in which some substances loses electrons to another substance, the oxidizing agent.

FOR FURTHER INFORMATION

Cornell, Rochelle M., and Udo Schwertmann. *The Iron Oxides; Structure, Properties, Reactions, and Uses.* Second edition. New York: Wiley-VCH, 2003.

"Ferrous Oxide." International Programme for Chemical Safety. http://www.inchem.org/documents/icsc/icsc/eics0793.htm (accessed on October 12, 2005).

"Material Safety Data Sheet." ESPI Metals. http://www.espimetals.com/msds's/ironoxidefeo.pdf (accessed on October 12, 2005).

See Also Iron(III) Oxide

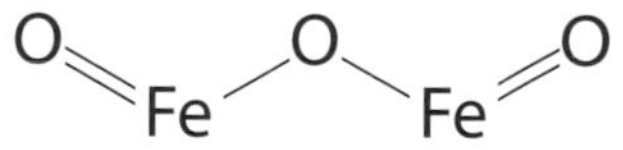

$$O=Fe-O-Fe=O$$

OTHER NAMES:
Ferric oxide; red iron
oxide; red iron
trioxide

FORMULA:
Fe_2O_3

ELEMENTS:
Iron; oxygen

COMPOUND TYPE:
Metallic oxide

STATE:
Solid

MOLECULAR WEIGHT:
159.69 g/mol

MELTING POINT:
1,565°C (2,849°F)

BOILING POINT:
Not applicable

SOLUBILITY:
Insoluble in water and
all conventional
organic solvents;
soluble in acids

K E Y   F A C T S

Iron(III) Oxide

OVERVIEW

Iron(III) oxide (EYE-urn three OK-side) is a dense, red-dish-brown, crystalline compound that usually occurs as lumps or a powder. It occurs in nature as the mineral hema-tite and is a component of the rust that forms on objects made out of iron that are exposed to the air. Rust itself is actually a complex mixture of iron oxides and hydroxides, including Fe_2O_3, FeO, Fe_3O_4 and $FeO(OH)$. Hematite may range in color from black and silver gray to reddish brown and red depending on the type and amount of impurities present with iron(III) oxide. Iron(III) is also ferromagnetic. Ferromagnetism refers to the ability of a substance to become highly magnetic and then retain its magnetism.

HOW IT IS MADE

Hematitite forms naturally when iron-containing rocks and minerals react with oxygen in the air to form iron(III) oxide. The oxide can be made synthetically by a variety of

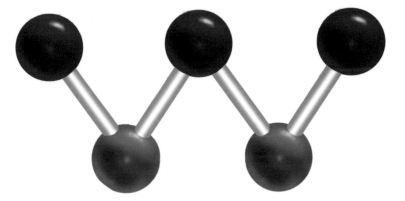

Iron(III) oxide. Red atoms are oxygen and orange atoms are iron. PUBLISHERS RESOURCE GROUP

procedures. In the most popular method, iron(II) sulfate is reacted with sodium hydroxide (NaOH) to produce iron(II) hydroxide [$Fe(OH)_2$]. The iron(II) hydroxide is then allowed to react with oxygen in the air, forming iron(III) oxide. The compound can also be produced by heating iron(II) sulfate, hydrated iron(II) oxide (FeO(OH)), or iron(III) oxalate [$Fe_2(C_2O_3)_3$].

COMMON USES AND POTENTIAL HAZARDS

Iron(III) oxides has been associated with the manufacture of iron and steel for much of human history. The Iron Age, which began in Egypt around 4000 BCE was the period in human history when iron was used for tools and weapons. The general approach to refining iron metal from iron ores, such as hematite, was to heat the ore in the presence of carbon. Carbon removes oxygen from the ore, leaving the free metal behind. By the first century BCE in China, the first known blast furnaces were in use. In a blast furnace, iron(III) oxide is reduced with carbon by using a blast of air and heat. The oxygen from the air reacts with carbon to give carbon monoxide, which then reacts with iron(III) oxide to produce liquid iron metal and carbon dioxide.

In the eighteenth century, the blast furnace process was further developed so that iron could be made commercially. This process can be traced to the region around

Interesting Facts

- The name hematite is derived from the Greek word for blood.

- When hematite is made into an ornament, it is sometimes called black diamond.

- NASA's Mars rover *Opportunity* found small particles thought to be mostly hematite on the planet's surface. Scientists think they formed billions of years ago when Mars had water on its surface.

- Paleolithic humans in Swaziland, who lived more than 40,000 years ago, mined hematite in the oldest known mine in the archaeological record called the Lion Cave. It is thought that they mined the hematite to produce the red pigment known as ochre.

Coalbrookdale in Shropshire, England, around the year 1773 and is said to have been a factor in initiating the Industrial Revolution. The blast furnace method is still one of the primary methods by which iron metal is refined from iron ores.

Iron(III) oxide is also one of the oldest known pigments and has been used for that purpose in every major civilization. Some of the best known pigments made from iron(III) oxide have been Indian red, terra Pozzuoli, and Venetian red and have been used to color ceramic glazes and paints. Depending on the exact formulation used, iron(III) oxide produces colors ranging from yellow to orange to red. For example, the hydrated oxide produces a pigment ranging from yellow to brown. Iron(III) oxide pigments have been used as pigment for rubber, paper, linoleum, glass, and many types of paints, including specialty paints used on metalwork and ship hulls.

Some of the other commercial and industrial applications of iron(III) oxide include:

Words to Know

CATALYST A material that increases the rate of a chemical reaction without undergoing any change in its own chemical structure.

MORDANT A substance used in dyeing and printing that reacts chemically with both a dye and the material being dyed to help hold the dye permanently to the material.

- As a catalyst for many industrial and chemical operations;

- As a component of thermite, a mixture of iron(III) oxide and aluminum powder which, when ignited, produces very hot temperatures. Thermite bombs are used in welding;

- In computer hard disks, audio cassette tapes, video cassette tapes, and computer floppy disks for magnetic storage of data;

- As an abrasive and polish for use with brass, steel, gems, and other hard objects;

- As a mordant in the dyeing of cloth;

- As a feed additive for domestic animals to ensure proper levels of iron in their systems; and

- In the manufacture of magnets and magnetic materials.

Exposure to iron(III) oxide dust can cause irritation of the eyes and throat. Long-term exposure to dust particles can cause chronic inflammation of the lungs.

FOR FURTHER INFORMATION

Crutchfield, Charlie. "Re: What Do the Different Colors of Rust Mean, Chemically." MadSci Network. http://www.madsci.org/posts/archives/2002-03/1015309769.Ch.r.html (accessed on October 12, 2005).

"Ferric Oxide." International Labour Organization. http://www.ilo.org/public/english/protection/safework/cis/products/icsc/dtasht/_icsc15/icsc1577.htm (accessed on October 12, 2005).

Ricketts, John A. "How a Blast Furnace Works." American Iron and Steel Institute.
http://www.steel.org/AM/Template.cfm?Section=Home& template=/CM/HTMLDisplay.cfm&ContentID=5433 (accessed on October 12, 2005).

See Also Carbon Monoxide; Iron(II) Oxide

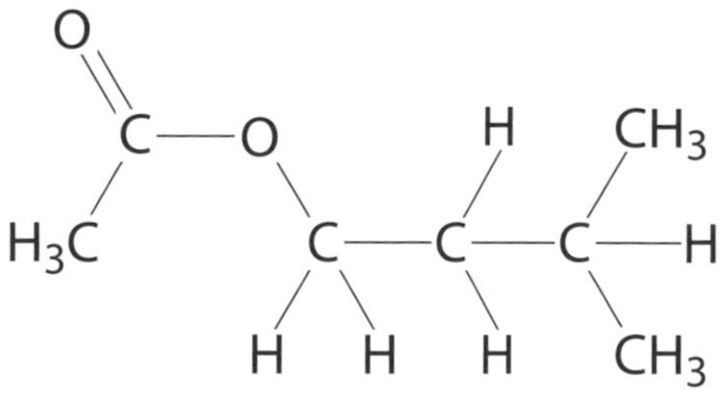

$$O$$
$$\|$$
$$H_3C - C - O - C - C - C - H$$

(structural diagram labels: O double bonded to C; H₃C; O; C with H, H; C with H and CH₃; C with H, CH₃, H)

OTHER NAMES:
Isopentyl acetate;
isoamyl ethanoate;
amylacetic ester

FORMULA:
$CH_3COOCH_2CH_2CH$
$(CH_3)_2$

ELEMENTS:
Carbon, hydrogen,
oxygen

COMPOUND TYPE:
Ester (organic)

STATE:
Liquid

MOLECULAR WEIGHT:
130.18 g/mol

MELTING POINT:
−78.5°C (−109°F)

BOILING POINT:
142.5°C (288.5°F)

SOLUBILITY:
Slightly soluble in
water; soluble in most
organic solvents

K E Y   F A C T S

Isoamyl Acetate

OVERVIEW

Isoamyl acetate (EYE-so-A-mil AS-uh-tate) is a clear, colorless liquid with a pleasant fruity odor and taste reminiscent of pears or bananas. When prepared for industrial or commercial use, it is often known as pear oil or banana oil.

HOW IT IS MADE

Isoamyl acetate is made commercially by reacting acetic acid (CH_3COOH) with amyl alcohol ($C_4H_9CH_2OH$) to produce amyl acetate, of which there are eight isomers. The isomers are then separated from each other by fractional distillation to obtain the one desired form, isoamyl acetate.

COMMON USES AND POTENTIAL HAZARDS

Isoamyl acetate is a very popular additive for imparting a pleasant odor or taste to commercial products. Since 1976, the U.S. Patent Office has issued 1,174 patents for inventions

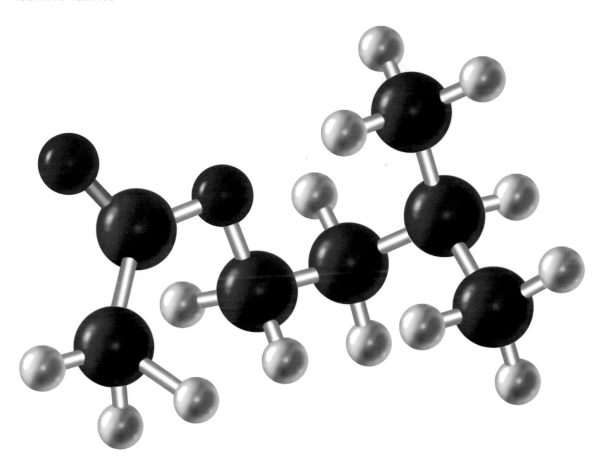

Isoamyl acetate. Red atoms are oxygen; white atoms are hydrogen; and black atoms are carbon. Gray sticks indicate double bonds.
PUBLISHERS RESOURCE GROUP

that contain the compound. Some of the products to which it is added include:

- Foods and drinks, such as gum, candy, mineral water, beer, and syrups used to make soft drinks;

- Personal care products, such as perfume, nail polish, leather polish, and shoe polish;

- Furniture polish, varnishes, and lacquers; and

- Dry cleaning preparation.

In addition to these applications, isoamyl acetate has a number of other uses, including

- The fermentation of grain to produce whiskey;

- To mask unpleasant odors;

- In the manufacture of a number of industrial and household products, including bath sponges, artificial

Interesting Facts

- When a bee stings, it leaves behind traces of isoamyl acetate at the site of the sting. The isoamyl acetate then attracts other bees to the same site, accounting for the tendency for an individual to receive multiple stings at the same point on his or her body.

- One species of Japanese honeybees defends itself from attacks by hornet predators by surrounding the hornet with a ball that consists primarily of isoamyl acetate. The ball becomes so hot that the hornet dies.

leather, artificial silk, rayon, artificial pearls, artificial glass, bronzing fluid, metallic paint, fluorescent lamps, and photographic film;

- In the dyeing and finishing of textiles; and

- As a solvent for old oil paints.

Isoamyl acetate is flammable and is rated as a severe fire hazard. It is also explosive. The chemical should not be used around open flames or sparks. People should not smoke around isoamyl acetate.

Isoamyl acetate is also an irritant to the skin, eyes, and respiratory and digestive systems. If swallowed it may cause sore throat, nausea, and abdominal pain. People who work

Words to Know

DISTILLATION A process of separating two or more substances by boiling the mixture of which they are composed and condensing the vapors produced at different temperatures.

ISOMER One of two or more forms of a chemical compound with the same molecular formula, but different structural formulas and different chemical and physical properties.

with the pure compound are at greater risk for harm from the compound than are those who use it in commercial products.

FOR FURTHER INFORMATION

"Isoamyl Acetate." International Programme on Chemical Safety. http://www.inchem.org/documents/icsc/icsc/eics0356.htm (accessed on October 12, 2005).

"Occupational Safety and Health Guideline for Isoamyl Acetate." Occupational Safety & Health Administration. http://www.osha.gov/SLTC/healthguidelines/isoamylacetate/recognition.html (accessed on October 12, 2005).

Shukla, Nutan. "Honeybees Come to Know of Queen's Death through Smell." *The Tribune Spectrum.* http://www.tribuneindia.com/2003/20030601/spectrum/nature.htm (accessed on October 12, 2005).

"Unusual Thermal Defence by a Honeybee against Mass Attack by Hornets." *Nature* (September 28, 1995): 334-336.

OTHER NAMES:
2-methyl-
1,3-butadiene

FORMULA:
$CH_2=CH(CH_3)CH=CH_2$

ELEMENTS:
Carbon, hydrogen

COMPOUND TYPE:
Alkadiene;
unsaturated
hydrocarbon
(organic)

STATE:
Liquid

MOLECULAR WEIGHT:
68.12 g/mol

MELTING POINT:
−145.9°C (−230.6°F)

BOILING POINT:
34.0°C (93.2°F)

SOLUBILITY:
Insoluble in water;
miscible with ethyl
alcohol, acetone,
ether, and benzene

KEY FACTS

Isoprene

OVERVIEW

Isoprene (EYE-so-preen) is a clear, colorless, volatile liquid that is both very flammable and quite explosive. It is classified as a *diene* compound because its molecules contain two ("di-") double bonds ("-ene"). It is also a member of the terpene family. The terpenes are a large family of organic compounds that contain two or more isoprene units. An example of a terpene is vitamin A, whose molecular formula is $C_{20}H_{30}O$. Vitamin A contains four isoprene units. The terpenes occur abundantly in nature in both plants and animals.

Some common terpenes include geraniol, found in geraniums; limonene, oil of orange; a-pinene, or oil of turpentine; a-farnesene, oil of cintronella; zingiberene, oil of ginger; farnesol, found in lily of the valley; β-selinene, oil of celery; and caryophyllene, oil of cloves. Isoprene is also produced in animal bodies and is said to be the most common hydrocarbon present in the human body. By one estimate, a 70-kilogram (150-pound) person produces about 17 milligrams of

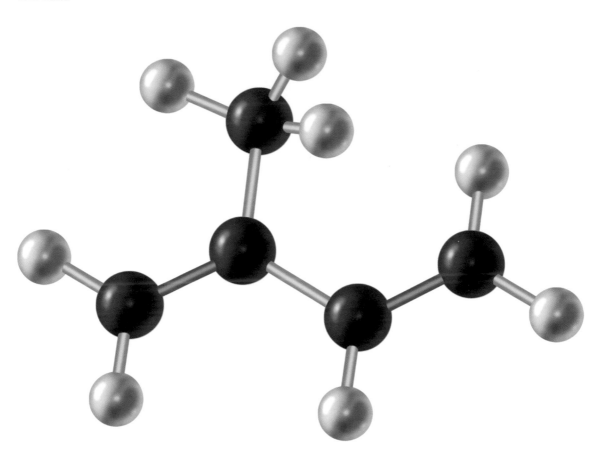

Isoprene. Black atoms are carbon; white atoms are hydrogen. Gray sticks show double bonds; white sticks show single bonds. PUBLISHERS RESOURCE GROUP

isoprene per day. Probably the best-known source of isoprene is natural rubber, which is a polymer consisting of long chains of isoprene units joined to each other.

HOW IT IS MADE

A number of methods are available for preparing isoprene from petroleum. Perhaps the most common process is the cracking of hydrocarbons present in the naphtha portion of refined petroleum. Cracking is the process by which large hydrocarbons are broken down into smaller hydrocarbons either with heat or over a catalyst, or by some combination of heat and catalyst. The naphtha portion of petroleum consists of hydrocarbons with boiling points between about 50°C and 200°C (120°F and 400°F). Other methods for the

Interesting Facts

- Isoprene and other terpenes are now known to undergo reactions that contribute to the development of pollutants, such as ozone and oxides of nitrogen in the atmosphere.

- Isoprene is a key intermediary in the synthesis of cholesterol in the human body.

- The production of isoprene by plants seems to be associated with the process of photosynthesis and is affected by temperature, sunlight, other gases, and other factors.

- The polymer of isoprene is called polyisoprene. It exists in two forms, *cis*- and *trans*-polyisoprene. The two forms are called *geometric isomers*. They have the same kind and number of atoms, but the atoms are arranged differently in the two forms. Natural rubber consists of *trans*-polyisoprene, while another product found in rubber plants, gutta percha, is made of *cis*-polyisoprene.

preparation of isoprene include the dehydrogenation (removal of hydrogen) of isopentene ($CH_3CH(CH_3)CH=CH_2$), the pyrolysis (decomposition by high heat) of methylpentene ($CH_2=C(CH_3)CH_2CH_2CH_3$), or the dehydration (removal of water) of methylbutenol ($CH_3C(CH_3)(OH)CH_2CH_3$).

COMMON USES AND POTENTIAL HAZARDS

Natural rubber has been known to humans for hundreds of years. Archaeologists have found that the Indians of South and Central America were making rubber products as early as the eleventh century. Until the end of the nineteenth century, natural supplies of rubber obtained from the rubber tree, *Hevea brasiliensis*, were sufficient to meet consumer demand for the product. However, with the development of modern technology—especially the invention of the automobile—natural supplies of the product proved to be insufficient to meet growing demand. Chemical researchers began to look for ways of producing synthetic forms of rubber.

One approach was to attempt making synthetic rubber with exactly the same chemical composition as that of natural rubber, that is, a polymer of *trans*-polyisoprene. As early as the 1880s, British chemist Sir William Augustus Tilden (1842-1926) was successful in achieving this objective. Tilden found that he could make isoprene by heating turpentine ($C_{10}H_{16}$). The isoprene then polymerized easily when exposed to light. After more than twenty years of research, however, Tilden decided that synthetic *trans*-polyisoprene could never be made economically, and he encouraged his friends to forget about the process.

Over the years, chemists did find ways of making other types of synthetic rubber, and some never abandoned the effort to make synthetic *trans*-polyisoprene. The critical breakthrough needed in this research occurred in about 1953 when Swiss chemist Karl Ziegler (1898-1973) and Italian chemist Giulio Natta (1903-1979) each found a way of polymerizing isoprene in such a way that its geometric structure matched that of natural rubber exactly. A year later, chemists at two of the largest rubber companies in the world, B. F. Goodrich and Firestone, announced that they had developed methods for making synthetic *trans*-polyisoprene using essentially the methods developed earlier by Ziegler and Natta.

In the early twenty-first century, more than 95 percent of the isoprene produced is used to make *trans*-polyisoprene synthetic rubber. The remaining 5 percent is used to make other types of synthetic rubber and other kinds of polymers. A small amount of the compound is used as a chemical intermediary, a substance from which other organic chemicals is made.

Isoprene is a dangerous fire hazard. It also poses a risk to human health and that of other animals. It is an irritant to skin, eyes, and the respiratory system. Upon exposure, it produces symptoms such as redness, watering, and itching of the eyes and itching, reddening, and blistering of the skin. If inhaled, it can irritate the lungs and respiratory system. Isoprene is a known carcinogen.

Words to Know

CARCINOGEN A chemical that causes cancer in humans or other animals.

MISCIBLE Able to be mixed; especially applies to the mixing of one liquid with another.

PHOTOSYNTHESIS The process by which green plants and some other organisms use the energy in sunlight to convert

carbon dioxide and water into carbohydrates and oxygen.

POLYMER A compound consisting of very large molecules made of one or two small repeated units called monomers.

VOLATILE Able to turn to vapor easily at a relatively low temperature.

FOR FURTHER INFORMATION

"Hazardous Substance Fact Sheet: Isoprene." New Jersey Department of Health and Senior Services. http://www.state.nj.us/health/eoh/rtkweb/1069.pdf (accessed on December 29, 2005).

"Isoprene." Shell Chemicals. http://www.shellchemicals.com/isoprene/1,1098,1116,00.html (accessed on December 29, 2005).

"Material Safety Data Sheet: Isoprene MSDS." ScienceLab.com. http://www.sciencelab.com/xMSDS-Isoprene-9924409 (accessed on December 29, 2005).

"United States Synthetic Rubber Program, 1939–1945." National Historic Chemical Landmarks, American Chemical Society. http://acswebcontent.acs.org/landmarks/landmarks/rbb/rbb_begin.html (accessed on December 29, 2005).

Weissermel, Klaus, and Hans-Jürgen Arpe. *Industrial Organic Chemistry.* Weinheim, Germany: Wiley-VCH, 2003, 117-122.

See Also Poly(Styrene-Butadiene-Styrene)

$$CH_3 - \underset{\underset{H}{|}}{\overset{}{C}} - CH_3$$
HO

Isopropyl Alcohol

KEY FACTS

OTHER NAMES:
2-propanol;
isopropanol; rubbing
alcohol

FORMULA:
$CH_3CHOHCH_3$

ELEMENTS:
Carbon, hydrogen,
oxygen

COMPOUND TYPE:
Alcohol (organic)

STATE:
Liquid

MOLECULAR WEIGHT:
60.10 g/mol

MELTING POINT:
−87.9°C (−126°F)

BOILING POINT:
82.3°C (180°F)

SOLUBILITY:
Miscible with water
and most common
organic solvents

OVERVIEW

Isopropyl alcohol (EYE-so-PRO-pil AL-ko-hol) is a colorless flammable liquid with a sweet odor. In 2004, about 600 million kilograms (about 1.3 billion pounds) of isopropyl alcohol were produced in the United States, with about half of that used as an industrial solvent and about a third used in the preparation of other chemical compounds. It is perhaps best known to many people as rubbing alcohol, usually a 70 percent solution of isopropyl alcohol in water. The compound is commonly used to clean a person's skin before an injection is given. It kills bacteria on the skin and prevents infection.

HOW IT IS MADE

The most popular industrial method for preparing isopropyl alcohol was invented in 1920 by researchers at the Standard Oil Company (now Exxon). In that process, propene (propylene; $CH_2CH{=}CH_2$) is treated with hydrolyzed with sulfuric acid as a catalyst.

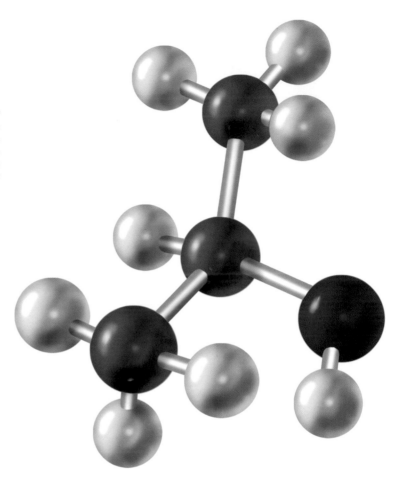

Isopropyl alcohol. Red atom is oxygen; white atoms are hydrogen; and black atoms are carbon. PUBLISHERS RESOURCE GROUP

COMMON USES AND POTENTIAL HAZARDS

Isopropyl alcohol dissolves many other organic compounds easily, so it finds wide use as a solvent for gums; essential and other kinds of oils; alkaloids; certain types of plastics; derivatives of cellulose; paints, varnishes, shellacs, and other types of coatings; and quick-drying inks. Essential oils are oils extracted from plants that have therapeutic value. Alkaloids are organic bases that contain the element nitrogen.

The synthesis of many important organic compounds begins with isopropyl alcohol as a raw material. Among these compounds are acetone, glycerol, and isopropyl acetate, itself widely used as a solvent for organic substances. Among the other uses to which isopropyl alcohol is put are:

Words to Know

HYDROLYSIS The process by which a compound reacts with water to form two new compounds.

MISCIBLE Able to be mixed; especially applies to the mixing of one liquid with another.

- In household and personal care products, such as perfumes, hair dye rinses, nail polishes, shampoos, and after-shave lotions;
- In cleaning products, such as disinfectant soaps and hand and body lotions;
- In antifreezes and as a deicing agent;
- As a coolant in the manufacture of beers; and
- As a preservative for biological specimens.

FOR FURTHER INFORMATION

"Isopropanol." Spectrum Laboratories. http://www.speclab.com/compound/c67630.htm (accessed on October 12, 2005).

"Isopropyl Alcohol." New Jersey Department of Health and Senior Services. http://www.state.nj.us/health/eoh/rtkweb/1076.pdf (accessed on October 12, 2005).

"2-Propanol." International Programme on Chemical Safety. http://www.inchem.org/documents/sids/sids/67630.pdf (accessed on October 12, 2005).

See Also Propylene

OTHER NAMES:
2-hydroxypropanoic acid; α-hydroxypropanoic acid; milk acid

FORMULA:
CH₃CHOHCOOH

ELEMENTS:
Carbon, hydrogen, oxygen

COMPOUND TYPE:
Carboxylic acid (organic)

STATE:
Liquid

MOLECULAR WEIGHT:
90.08 g/mol

MELTING POINT:
16.8°C (62.2°F)

BOILING POINT:
Not applicable; decomposes upon heating

SOLUBILITY:
Very soluble in water and ethanol; slightly soluble in ether

KEY FACTS

Lactic Acid

OVERVIEW

Lactic acid (LAK-tik AS-id) is a colorless, odorless, syrupy liquid that occurs in two isomeric forms, D-lactic acid and L-lactic acid. Isomers are two or more forms of a chemical compound with the same molecular formula, but different structural formulas and different chemical and physical properties. The D form is produced during metabolic reactions that take place in muscle tissue, while the L form is produced by yeast cells. The synthetic production of lactic acid results in a product consisting of equal amounts of the D and L forms, a mixture known as a racemic mixture.

Lactic acid was first discovered in 1780 by the Swedish chemist Karl Wilhelm Scheele (1742-1786), who called his discovery "acid of milk." The two isomeric forms of the acid were first identified in 1863 by the German chemist Johannes Wislicenus (1835-1902), and the compound was first produced commercially in 1881 by American chemist Charles E. Avery. Avery patented his invention in 1885 and constructed a factory for the production of lactic acid in Littleton, Massachusetts.

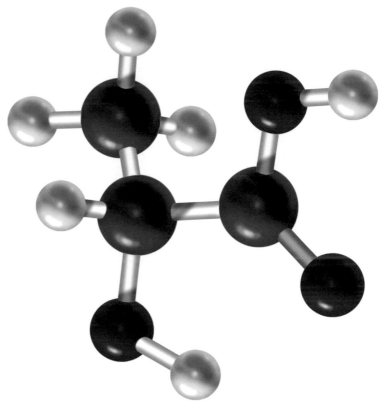

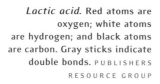

Lactic acid. Red atoms are oxygen; white atoms are hydrogen; and black atoms are carbon. Gray sticks indicate double bonds. PUBLISHERS RESOURCE GROUP

About 30 million kilograms (72 million pounds) of lactic acid are produced annually in the United States. The most common method of production is the fermentation of glucose by yeast.

HOW IT IS MADE

In muscle cells, lactic acid is the product of anaerobic respiration, the process by which glucose is oxidized in the absence of oxygen to produce energy required by cells. Although some lactic acid is always produced in muscle cells in very low concentrations, it tends to accumulate during exercise, when cells do not receive adequate amounts of oxygen to metabolize oxygen by normal pathways. Lactic acid produced during exercise remains in the body for only short periods of time, sometimes in less than thirty minutes. It is metabolized in the muscle cells where it was produced, resulting in the production of energy, carbon dioxide, water, and other products.

Interesting Facts

- For the better part of a century, athletes and physiologists have considered lactic acid a primary cause of fatigue during high-intensity exercise. However, scientists have learned that lactic acid actually helps to prevent muscle fatigue. Muscle soreness once thought to be caused by lactic acid is, instead, more likely to be a result of damaged muscle cells caused by excess use.

- Lactic acid present on the skin attracts mosquitoes.

- Lactic acid in the body exists in its ionic form, known as lactate.

Lactic acid is also produced by yeast during the process of fermentation. Fermentation is the process by which yeast cells convert glucose to an alcohol and carbon dioxide. Yeast cells use almost precisely the same enzyme in fermentation that muscle cells use in anaerobic respiration. The muscle cell enzyme and the yeast enzyme differ only in the orientation of one group of atoms, resulting in the production of the D isomer in one case and the L isomer in the other.

A synthetic process for the production of lactic acid was first introduced in 1963. That process begins with the addition of hydrogen cyanide (HCN) to acetaldehyde (ethanal; CH_3CHO), resulting in the formation of lactonitrile (CH_3CH_2OCN). The lactonitrile is then hydrolyzed, using a strong acid, such as sulfuric acid, as a catalyst, to make lactic acid.

COMMON USES AND POTENTIAL HAZARDS

The primary use for lactic acid in the United States is as a food additive, where it acts as an acidulant and a flavor additive. An acidulant is a compound that provides an acidic environment for foods, as is the case with yogurt, buttermilk, sauerkraut, green olives, pickles, and other acidic foods. As a flavor additive, it adds a tart or tangy flavor to foods and beverages, as well as acting as a preservative to keep them from spoiling. Lactic acid also has a number of important industrial uses, the most important of which is the

Words to Know

CATALYST A material that increases the rate of a chemical reaction without undergoing any change in its own chemical structure.

ELECTROPLATING Adding a layer of nickel, silver, or gold, on another type of metal using an electric current.

FERMENTATION The process by which yeast convert glucose to an alcohol and carbon dioxide.

HYDROLYSIS The process by which a compound reacts with water to form two new compounds.

ISOMER One of two or more forms of a chemical compound with the same molecular formula, but different structural formulas and different chemical and physical properties.

METABOLISM A process that includes all of the chemical reactions that occur in cells by which fats, carbohydrates, and other compounds are broken down to produce energy and the compounds needed to build new cells and tissues.

MORDANT A substance used in dyeing and printing that reacts chemically with both a dye and the material being dyed to help hold the dye permanently to the material.

SYNTHESIS A chemical reaction in which some desired chemical product is made from simple beginning chemicals, or reactants.

production of other organic chemicals, especially ethyl lactate, acrylic acid, propylene glycol, and the polymer known as polyactide. Polyactide is used in the manufacture of plastic film, fiber, packaging material, and filling materials. Other commercial and industrial applications of lactic acid include:

- As a mordant in dyeing;
- As a solvent for dyes that are not soluble in water;
- For the treatment of animal hides in the preparation of leather products;
- As a catalyst in the production of certain types of plastics; and
- As an additive in electroplating baths.

Lactic acid in normal concentrations poses no safety or health hazards to humans or other animals. One health consequence related to lactic acid, however, is a condition known as gout, a type of arthritis that causes severe pain in the joints. Gout is caused by an accumulation of uric acid in the blood. Since lactic acid blocks the elimination of uric acid from the body, individuals with excess lactic acid buildup, usually caused by high alcohol consumption, may develop an excess of uric acid crystals in the blood and joints, leading to gout.

FOR FURTHER INFORMATION

"Cell Respiration." SparkNotes. http://www.sparknotes.com/testprep/books/sat2/biology/ chapter6section1.rhtml (accessed on October 14, 2005).

Drake, Geoff. "The Lactate Shuttle—Contrary to What You've Heard, Lactic Acid Is Your Friend." *Bicycling* (August 1992): 36.

Friel, Joel. "All Athletes: Lactic Acid's Bad Rap." Ultrafit's e-Tips For Endurance Athletes. October 2004, Vol. 7, No. 10. http://www.ultrafit.com/newsletter/october04.html#Joe (accessed on October 14, 2005).

"Lactic Acid." J. T. Baker. http://www.jtbaker.com/msds/englishhtml/l0522.htm (accessed on October 14, 2005).

Rogers, Palmer, Jiann-Shin Chen, and Mary Jo Zidwick. *Organic Acid and Solvent Production, Part I: Acetic, Lactic, Gluconic, Succinic, and Polyhydroxyalkanoic Acids.* Section 2: Lactic Acid. Available online at http://141.150.157.117:8080/prokPUB/chaphtm/306/ 04_00.htm (accessed on October 14, 2005).

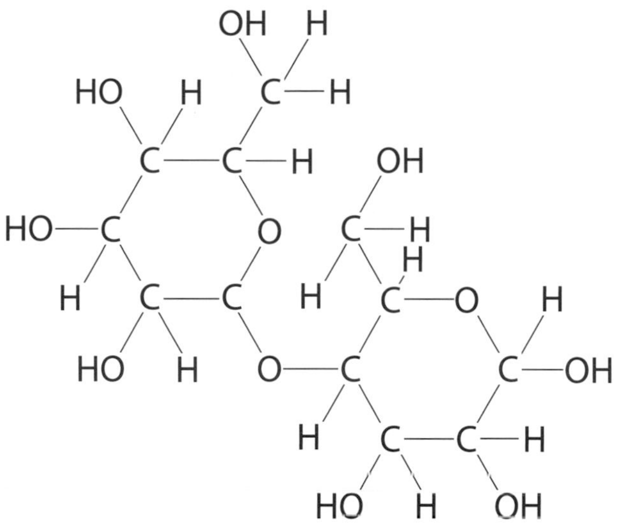

OTHER NAMES:
D-lactose; milk sugar;
many others

FORMULA:
$C_{12}H_{22}O_{11}$

ELEMENTS:
Carbon, hydrogen,
oxygen

COMPOUND TYPE:
Disaccharide;
carbohydrate
(organic)

STATE:
Solid

MOLECULAR WEIGHT:
342.30 g/mol

MELTING POINT:
222.8°C (433.0°F)

BOILING POINT:
Not applicable;
decomposes

SOLUBILITY:
Very soluble in water;
slightly soluble
in ethyl alcohol;
insoluble in organic
solvents

KEY FACTS

Lactose

OVERVIEW

Lactose (LAK-tose) is a white, odorless, sweet-tasting solid commonly known as milk sugar because it occurs in the milk of many animals, primarily the mammals. The lactose content of milk ranges from about 2 to 8 percent in cows and 5 to 8 percent in human milk. Lactose occurs in two isomeric forms, α-lactose and β-lactose, with the latter somewhat sweeter than the former. The alpha form tends to occur as the monohydrate, $C_{12}H_{22}O_{11} \cdot H_2O$.

HOW IT IS MADE

Lactose is synthesized in the mammary (milk-producing) glands of mammals. The milk of such animals contains an enzyme called lactose synthetase, which acts on the compound uridine diphosphate D-galactose to produce lactose. The compound is obtained commercially from whey, a byproduct of the cheese-making process. The solids in whey contain about 70 percent lactose by weight. These solids are

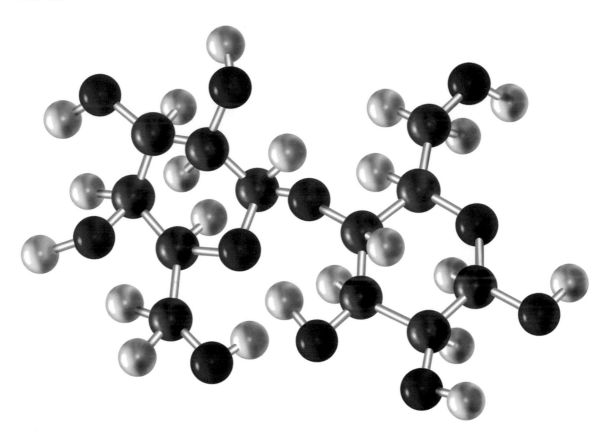

Lactose. Red atoms are oxygen; white atoms are hydrogen; and black atoms are carbon. PUBLISHERS RESOURCE GROUP

filtered to remove the proteins they contain. After removal of minerals in the whey, the resulting solution consists of about 50 to 65 percent by weight, which is allowed to crystallize out of the resulting solutions. The lactose produced by this process is a racemic mixture of the D and L isomers. To obtain the alpha isomer, the lactose is dissolved with water and treated with activated carbon to remove any color. When the water evaporates from the solution, α-lactose monohydrate remains. The product is most commonly made available in this form. To obtain the beta isomer, α-lactose is heated with water in the presence of a base, which converts the alpha isomer to its beta form.

COMMON USES AND POTENTIAL HAZARDS

Both forms of lactose are used in the preparation of baby foods and food for infants and for convalescents. It is also used in the dairy industry to feed foals that have been

Interesting Facts

- Only one group of mammals does not produce lactose in their milk, the Pinnipedia, a group that includes seals, sea lions, and walruses.

- Alpha lactose is found in cow's milk and beta lactose in human milk. During pasteurization of cow's milk, the alpha isomer is converted into the beta isomer.

orphaned. Lactose is also used as a food additive, primarily as a humectant and to increase the sweet flavor of a food. A humectant helps foods retain their moisture. It is used in products such as ice creams, baked goods, confectionary items, whipped toppings, and breakfast foods. Lactose may also be added to a number of foods with limited natural sweetness, such as margarine, butter, frozen vegetables, and processed meats. It is also used as a filler for many pharmaceutical products. A filler acts bulk to the product without significantly changing its nutritional or medical properties.

Many people are lactose intolerant. Lactose intolerance is a condition that develops when a person lacks enough (or any) of the enzyme lactase that is responsible for digestion of lactose in the body. People who are lactose intolerant and consume lactose experience a range of unpleasant symptoms, including fluid retention, gas, cramps, and diarrhea. By some estimates, up to 75 percent of the world's population may experience lactose intolerance to a greater or lesser degree.

One method for dealing with lactose intolerance is for a person to avoid consumption of any food or food product that contains lactose. The option has been made somewhat easier by the introduction of lactose-free products by a number of food companies. Another method to avoid the problems associated with lactose intolerance is to use a dietary supplement that contains lactase, restoring to the body the enzyme that it lacks naturally.

Words to Know

HYDRATE A chemical compound formed when one or more molecules of water is added physically to the molecule of some other substance.

MAMMALS A group of warm-blood animals whose females produce milk for their young. This group includes humans, cows, bears, and dogs, among others.

RACEMIC MIXTURE A mixture of equal amounts of two opposite isomers (the "right-handed," or "D" form and the "left-handed," or "L" form).

FOR FURTHER INFORMATION

"Lactose." J. T. Baker.
http://www.jtbaker.com/msds/englishhtml/l1044.htm
(accessed on October 14, 2005).

"Lactose Intolerance." Medline Plus.
http://www.nlm.nih.gov/medlineplus/lactoseintolerance.html
(accessed on October 14, 2005).

"Lactose Intolerance." National Digestive Diseases Information Clearinghouse.
http://digestive.niddk.nih.gov/ddiseases/pubs/
lactoseintolerance/ (accessed on October 14, 2005).

See Also Sucrose

OTHER NAMES:
Aspartame

FORMULA:
$C_{14}H_{18}N_2O_5$

ELEMENTS:
Carbon, hydrogen,
nitrogen, oxygen

COMPOUND TYPE:
Ester (organic)

STATE:
Solid

MOLECULAR WEIGHT:
294.30 g/mol

MELTING POINT:
246.5°C (475.7°F)

BOILING POINT:
Not applicable

SOLUBILITY:
Slightly soluble in
water; insoluble in
alcohol, benzene,
ether, and most other
organic solvents

K E Y   F A C T S

L-Aspartyl-L-Phenyl-alanine Methyl Ester

OVERVIEW

L-aspartyl-L-phenylalanine methyl ester (ell-ass-par-TEEL ell-fee-no-AL-uh-neen METH-el ESS-ter) is an artificial sweetener more widely known as aspartame. It is sold under a number of brand names, including NutraSweet®, Equal®, Spoonful®, Benevia®, Indulge®, NatraTaste®, and Equal-Measure®. Unlike sugar, which is a carbohydrate, aspartame is a dipeptide, a compound made of two amino acids joined to each other. It is 180 to 200 times as sweet as sucrose (table sugar), but provides no calories to a person's diet. It is a satisfactory substitute for sugar, therefore, for people who must or wish to reduce their caloric intake.

Aspartame was discovered accidentally in 1965 by James M. Schlatter, an employee of the G. D. Searle pharmaceutical company. Schlatter was searching for a chemical compound that could be used to treat ulcers. L-aspartyl-L-phenylalanine methyl ester was one of the compounds he made during his investigations. He accidentally got some of the compound on his fingers, a fact that he did not notice until later in the day.

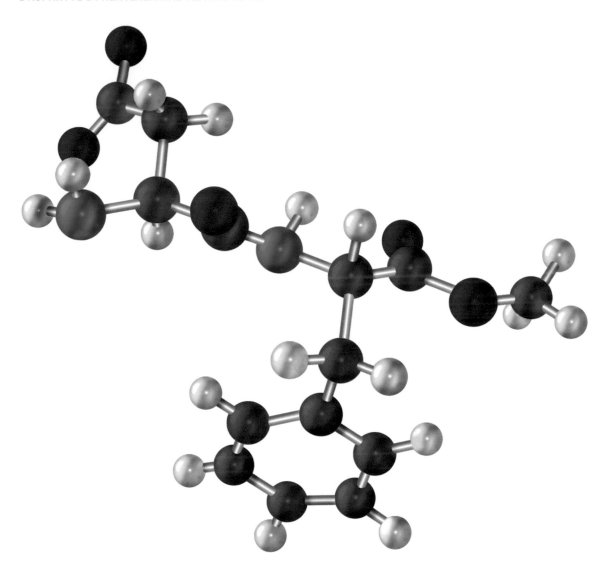

L-Aspartyl-L-phenylalanine methyl ester. Red atoms are oxygen; white atoms are hydrogen; black atoms are carbon; and blue atoms are nitrogen. Gray sticks show double bonds. PUBLISHERS RESOURCE GROUP

Then, when he licked his fingers to pick up a piece of paper, he noticed the intensely sweet taste of the compound. To confirm his suspicions about the compound, he added some of it to his coffee—a practice that violates all laboratory safety rules!—and found that it was more than satisfactory as a substitute for sugar. Searle began the testing and application procedure needed to gain approval from the U.S. Food and Drug Administration, a long and drawn-out procedure that ended only in 1981, 25 years after its discovery.

Interesting Facts

- The Aspartame Information Center claims that the compound is now used in more than 6,000 commercial products and consumed by more than 100 million people worldwide.

HOW IT IS MADE

Aspartame is made by a relatively simple procedure in which two amino acids, aspartic acid and phenylalanine, are reacted with each other to form a two-amino-acid product, called a dipeptide. The carboxylic acid group in the dipeptide is then reacted with methanol (methyl alcohol; CH_3OH) to obtain the methyl ester of the compound. One problem that makes the preparation somewhat more difficult is that both aspartic acid and phenylalanine have stereoisomers. The term stereoisomer refers to two forms of a compound that contain the same kind and number of atoms, but differ in the orientation in space ("stereo-") of some of the atoms. Because of these stereoisomers, four different kinds of aspartame are formed during the preparation described above – D & D; L & L; D & L; L & D (the latter two are different from each other). Only one is the desired product, the one that contains only "L" stereoisomers.

COMMON USES AND POTENTIAL HAZARDS

Aspartame's primary commercial use is as an artificial sweetener. Some of the products that contain aspartame are breath mints, carbonated soft drinks, cereals, chewing gum, sugar-free gelatin, hard candies, ice cream toppings, sugar-free ice cream, ready-to-drink iced tea, instant cocoa mix, jams and jellies, juice drinks, nutritional bars, protein nutritional drinks, diet puddings, sugar-free chocolate syrup, sugar-free cookies, table top sweeteners, and fat- and sugar-free yogurt. The compound is also used to a lesser extent as a flavor enhancer, a substance that intensifies an already-existing flavor.

The use of aspartame as a food additive in the United States has a long and complex history. As required for FDA approval, Searle conducted a number of studies to show that aspartame is safe for human consumption. Based on the results of those studies, the FDA first granted approval for the use of aspartame in certain types of foods on July 26, 1974. Less than a month later, two concerned citizens, James Turner and Dr. John Olney, filed a petition objecting to the FDA's decision, citing possible errors in Searle's testing procedures on aspartame. After continued studies extending over a seven-year period, the FDA once again approved aspartame for use in dry foods. A year later, the agency extended its approval for aspartame to dry beverage mixes and add-in sweeteners and, in 1983, to beverages and other foods.

The campaign against aspartame has not lessened over the years. Today, a number of individuals and groups maintain organizations and websites whose purpose it is to have aspartame banned as a food additive. Continued studies by the manufacturer and independent researchers seem to confirm the safety of aspartame, but have found that two groups of people do face some level of risk by ingesting the compound. The first group consists of people who seem to have some natural allergy to the compound and experience symptoms such as anxiety attacks, breathing difficulties, dizziness, unusual fatigue, headaches, heart palpitations, skin rashes, muscle spasms, nausea, numbness, respiratory allergies, weight gain, and memory loss. Individuals who experience such problems are encouraged to avoid eating or drinking products that contain aspartame.

A second group for whom aspartame is a far more serious problem consists of individuals with phenylketonuria. Phenylketonuria is a genetic disorder in which a person's body is unable to metabolize the amino acid phenylalanine, a component of aspartame. If phenylalanine is ingested by someone with phenylketonuria, the amino acid builds up in the body and causes a number of organic problems, including mental retardation, damage to the organs, and muscular disorders. Any product that contains aspartame is now required by law to have a warning label aimed at phenylketonuriacs so that they will not accidentally ingest the compound.

Words to Know

DIPEPTIDE A compound made of two amino acids joined to each other.

METHYL ESTER A compound formed when methyl alcohol (methanol) reacts with an organic acid.

PHENYLKETONURIA A genetic disorder in which a person's body is unable to metabolize the amino acid phenylalanine.

STEREOISOMER One of a pair of molecules, both of which have the same kinds and number of atoms, but which differ in how some of the atoms are oriented in relation to each other.

FOR FURTHER INFORMATION

"Aspartame." Chemical Land 21. http://www.chemicalland21.com/lifescience/foco/ASPARTAME.htm (accessed on September 21, 2005).

"Aspartame Information Center." http://www.aspartame.org/ (accessed on September 21, 2005).

"Low-Calorie Sweeteners: Aspartame." Calorie Control Council. http://www.caloriecontrol.org/aspartame.html (accessed on September 21, 2005).

Metcalfe, Ed., et al. "Sweet Talking." *The Ecologist* (June 2000): 16.

OTHER NAMES:
5-amino-2,3-dihydro-
1,4-phthalazinedione;
3-aminophthalhydra-
zide

FORMULA:
$C_8H_7N_3O_2$

ELEMENTS:
Carbon, hydrogen,
nitrogen, oxygen

COMPOUND TYPE:
Heterocyclic ring
(organic)

STATE:
Solid

MOLECULAR WEIGHT:
177.16 g/mol

MELTING POINT:
319°C to 320°C (606°F
to 608°F)

BOILING POINT:
Not applicable;
decomposes

SOLUBILITY:
Slightly soluble in
water; soluble in
alcohol

Luminol

OVERVIEW

Luminol (LOO-min-ol) is a substance that glows when it
come in contact with blood. It was discovered in the late
nineteenth century, but chemists found little use for the
compound for half a century. Then, in 1928, the German
chemist H. O. Albrecht found that the addition of hydrogen
peroxide to luminol produces a bluish-green glow, an example of the phenomenon known as chemiluminescence. Chemiluminescence is the process by which light is emitted as
the result of a chemical reaction. Albrecht found that the
reaction between hydrogen peroxide and luminol required a
catalyst, a small amount of a metal such as copper or iron.

The most important application for luminol was discovered in 1937 by the German forensic scientist Walter Specht
(1907-1977), at the University Institute for Legal Medicine
and Scientific Criminalistics in Jena, Germany. Specht
found that blood itself could act as the catalyst needed to
produce chemiluminescence with luminol. He simply
sprayed a mixture of luminol and hydrogen peroxide on a

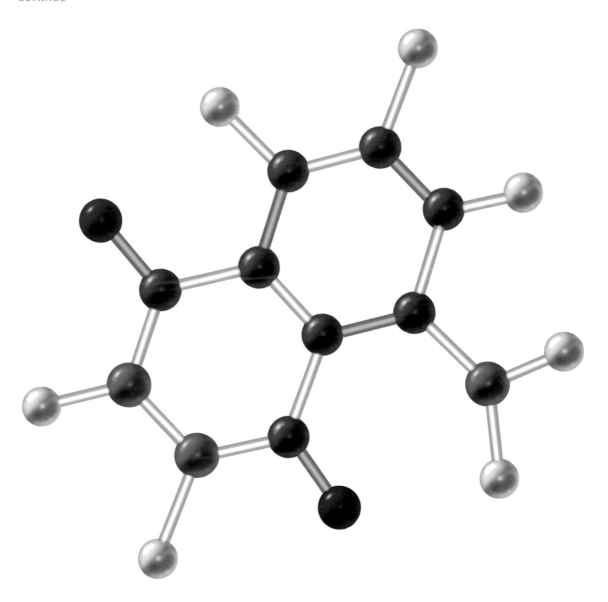

Luminol. Red atoms are oxygen; white atoms are hydrogen; black atoms are carbon; and blue atoms are nitrogen. Gray sticks indicate double bonds. PUBLISHERS RESOURCE GROUP

drop of blood, and the blood emitted a bluish-green glow. Specht later determined the explanation for this reaction. Blood contains a protein called hemoglobin that carries oxygen from the lungs to cells. Hemoglobin is a complex molecule with a single iron atom at its center. The small amount of iron in hemoglobin is sufficient to initiate the chemiluminescent reaction between luminol and hydrogen peroxide.

Interesting Facts

- Luminol can be used to detect bloodstains that are many years old.

- One disadvantage of using luminol in testing for blood is that it destroys the sample being investigated, making further tests on the same sample impossible.

HOW IT IS MADE

Luminol is prepared commercially by treating 3-nitrophthalic acid ($C_8H_5NO_6$) with hydrazine (NH_2NH_2), resulting in the formation of nitrophthalhydrazide. Nitrophthalhydrazide is then treated with sodium bisulfite ($NaHSO_3$) to obtain luminol.

COMMON USES AND POTENTIAL HAZARDS

The most common application of luminol is to find traces of blood at crime scenes. When a violent crime is committed, a certain amount of blood is often spilled on the floor, walls, furniture, and other objects at the crime scene. The perpetrator of the crime may attempt to clean up after the crime, but it is virtually impossible to remove all traces of blood. Forensic scientists who investigate a crime scene often assume that blood is present, even if it is not obvious. They check for the blood by spraying a mixture of luminol and hydrogen peroxide around the crime scene. If blood is present, it glows with a bluish-green color. The distribution of the blood can provide information as to where the crime was committed, whether the injured or murdered person was moved, and, if so, in what direction. Investigators typically take photographs of the illuminated crime scene for study at a later date.

False positive results are possible with a luminol test. A false positive test is a test in which the results seem to indicate the presence of blood even if it is not actually there. Some metals, plants, paints, cleaning materials, and other

Words to Know

CATALYST A material that increases the rate of a chemical reaction without undergoing any change in its own chemical structure.

substances may act as catalysts for the reaction between luminol and hydrogen peroxide and give a false positive test. For this reason, positive tests obtained by the luminol reagent are always subjected to further tests to confirm the results.

Luminol does have applications beyond criminal investigations. It is the active ingredient in glow sticks, the plastic sticks that glow green when broken. Chemists use the compound in chromatography, a process by which the components of a mixture are separated from each other, as well as in studies of DNA patterns and other biochemical tests.

FOR FURTHER INFORMATION

Genge, Ngaire E. *The Forensic Casebook: The Science of Crime Scene Investigation.* New York: Ballantine, 2002.

"Material Safety Data Sheet for Luminol = 3-Aminophthalhydrazide, 98%." Department of Chemistry, Iowa State University. http://avogadro.chem.iastate.edu/MSDS/luminol.htm (accessed on October 14, 2005).

"Nitric Oxide/NOS Detection (incl. Kits/Sets)." Axxora. http://www.axxora.com/nitric_oxide-nos_detection_(incl._kits-sets)-ALX-610-002/opfa.1.1.ALX-610-002.169.4.1.html (accessed on October 14, 2005).

"Technical Note: Hemaglow™." Lightning Powder Company. http://www.redwop.com/technotes.asp?ID=118 (accessed on October 14, 2005).

$$Cl^- \quad\quad Cl^-$$
$$Mg^{++}$$

Magnesium Chloride

OVERVIEW

Magnesium chloride (mag-NEE-zee-um KLOR-ide) is a white crystalline solid that is strongly deliquescent. It absorbs moisture from the air to become the hydrated form, magnesium chloride hexahydrate (MgCl$_2$·6H$_2$O). A deliquescent substance is one that takes on moisture from the air, often to the extent of dissolving in its own water of hydration. Magnesium chloride is an important industrial chemical, used in the production of magnesium, textile and paper manufacture, and cements; in refrigeration and fireproofing; and as a deicing agent.

HOW IT IS MADE

Magnesium chloride is extracted from seawater or brine, of which it is a component, and from minerals, such as carnallite (KCl·MgCl$_2$·H$_2$O) and bischofite (MgCl$_2$·6H$_2$O). The usual procedure is to treat seawater, brine, or the mineral with lime (CaO), calcined dolomite (CaO·MgO), or caustic soda (sodium

2+

Magnesium chloride hexahydrate. Green atoms are chlorine; and turquoise atom is magnesium. PUBLISHERS RESOURCE GROUP

hydroxide; NaOH) to make magnesium hydroxide [$Mg(OH)_2$], which is then treated with hydrochloric acid (HCl) to recover magnesium chloride, which is usually obtained as the crystalline hexahydrate. The pure compound is also produced by heating the double salt, magnesium ammonium chloride ($MgCl_2 \cdot NH_4Cl \cdot 6H_2O$), which first loses its water of hydration to form the anhydous double salt ($MgCl_2 \cdot NH_4Cl$). With further heating, the ammonium chloride sublimes, leaving behind pure anhydrous magnesium chloride.

COMMON USES AND POTENTIAL HAZARDS

The largest single use of magnesium chloride is in the production of magnesium metal. The metal is obtained through the electrolysis of molten magnesium chloride in a process developed by the American chemist and inventor Herbert Henry Dow (1866-1930) in 1916. The Dow process is still the primary method used for the production of magnesium metal. Other commercial and industrial uses of magnesium chloride include:

- In the manufacture of disinfectants;
- In the fireproofing of steel beams, wooden panels, and other materials;
- As a component of fire extinguishers;
- In the manufacture of so-called Sorel cement, a mixture of magnesium chloride and magnesium oxide, also known as oxychloride cement;
- As a binder to control dust on dirt roads;
- As a deicing compound;

Interesting Facts

- Tofu is traditionally prepared by treating soy milk with magnesium chloride or calcium sulfate.

- To remove suspended particles in water and sewage treatment plants;

- For the treatment of cotton and wool fabrics;

- In the processing of sugar beets;

- To keep drilling tools cool; and

- In the manufacture of paper and ceramic materials.

Magnesium chloride is a skin, nose, and eye irritant, although that hazard is usually a matter of concern only to those who work with the pure compound. It is also toxic by ingestion. Swallowing the compound can produce nausea, vomiting, and diarrhea. Inhalation of magnesium chloride fumes can irritate the lungs and respiratory tract, producing a condition known as metal fume fever that resembles the flu.

Words to Know

ANHYDROUS A compound that lacks any water of hydration.

DELIQUESCENT Describing a substance that takes on moisture from the air, often to the extent of dissolving in its own water of hydration.

HYDRATE A chemical compound formed when one or more molecules of water is added physically to the molecule of some other substance.

SUBLIMATION The process by which a solid changes directly into a gas without first melting.

WATER OF HYDRATION Water that has combined with a compound by some physical means.

FOR FURTHER INFORMATION

"Magnesium Chloride." ChemicalLand21.com.
    http://www.chemicalland21.com/arokorhi/industrialchem/
    inorganic/MAGNESIUM%20CHLORIDE.htm   (accessed   on
    October 14, 2005).

"Magnesium Chloride." J. T. Baker.
    http://www.jtbaker.com/msds/englishhtml/m0156.htm
    (accessed on October 14, 2005).

See Also Magnesium Hydroxide

H $\quad$ O⁻ $\quad$ Mg²⁺ $\quad$ O⁻ $\quad$ H

OTHER NAMES:
Magnesium hydrate;
milk of magnesia;
magnesia magma

FORMULA:
$Mg(OH)_2$

ELEMENTS:
Magnesium,
hydrogen, oxygen

COMPOUND TYPE:
Inorganic base

STATE:
Solid

MOLECULAR WEIGHT:
58.33 g/mol

MELTING POINT:
Decomposes at
350°C (660°F)

BOILING POINT:
Not applicable

SOLUBILITY:
Virtually insoluble in
water and alcohol,
soluble in dilute acids
and solutions of
ammonium salts

KEY FACTS

Magnesium Hydroxide

OVERVIEW

Magnesium hydroxide (mag-NEE-zee-um hye-DROK-side) is a white powder with no odor, found in nature as the mineral brucite. Perhaps the best known form of the compound is a milky liquid known as milk of magnesia, a product used to treat upset stomach and constipation. Milk of magnesia was invented in 1817 by the Irish pharmacist Sir James Murray (1788–1871). Murray built a plant to produce a mixture of magnesium hydroxide in water that he sold for the treatment of a variety of ailments, including heartburn, stomach acidity, bladder and bowel problems, and "female problems." He said that the liquid mixture was much more effective than powdery magnesium hydroxide which had previously been used for the same purposes.

In 1880, New York chemist Charles Henry Phillips (1820–1882) invented the name "milk of magnesia" and opened his own factory for producing the product. The name Phillips Milk of Magnesia is one of the oldest and best known over-the-counter medicines ever made in the United States.

2+

Magnesium hydroxide. Red atoms are oxygen; white atoms are hydrogen; and turquoise atom is magnesium.

HOW IT IS MADE

Magnesium hydroxide is prepared by reacting a magnesium salt, such as magnesium chloride ($MgCl_2$), with sodium hydroxide (NaOH). In a similar procedure, seawater (which contains small amounts of magnesium chloride) is treated with lime (calcium oxide; CaO). The water, lime, and magnesium chloride react to produce magnesium hydroxide, which settles out of solution as a precipitate.

COMMON USES AND POTENTIAL HAZARDS

The best known use for magnesium hydroxide is as an antacid in the form of milk of magnesia. Since magnesium hydroxide is a base, it reacts with excess stomach acid, which reduces heartburn and the discomfort of upset stomach.

Interesting Facts

- The element magnesium is named after a region in Greece called Magnesia in ancient times. Today the same region is called Manisa. Compounds of magnesium, including magnesium hydroxide, were abundant in the region and were given the name of *magnesia lithos*, or "stones of Magnesia."

Words to Know

ANTACID A medicine used for the treatment of upset stomach, acid indigestion, and related symptoms.

CLARIFIER A substance that removes impurities from another substance.

PRECIPITATE A solid material that settles out of a solution, often as the result of a chemical reaction.

SUSPENSION A mixture of two substances that do not dissolve in each other.

Milk of magnesia also acts as a laxative because it increases the flow of water into the intestines, stimulating a bowel movement. Milk of magnesia consists of an 8 percent suspension of solid magnesium hydroxide in water. Other chemicals, such as calcium carbonate and aluminum hydroxide, are sometimes added to the mixture to increase its effectiveness.

Magnesium hydroxide also has a number of important industrial uses, such as:

- A clarifier (a substance that removes impurities) in the refining of sugar;
- An additive in the treatment of wastewater to neutralize acids present in the wastes;
- A flame retardant coating on fabrics and other materials used by consumers and industries;
- An additive to fuel oils;
- An additive in toothpastes; and
- A drying agent in some food products.

FOR FURTHER INFORMATION

Dean, Carolyn. *The Miracle of Magnesium.* New York: Ballantine Books, 2003.

"Magnesium Hydroxide." Chemical Land 21. http://www.chemicalland21.com/arokorhi/industrialchem/inorganic/MAGNESIUM%20HYDROXIDE.htm (accessed on September 14, 2005).

See Also Magnesium Oxide

$$Mg = O$$

Magnesium Oxide

OVERVIEW

Magnesium oxide (mag-NEE-see-um OK-side) is available commercially in several forms, depending on the way it is prepared and the use for which it is intended. Most forms can be classified as either "light" or "heavy" depending on particle size, purity, and method of production. It occurs in nature in the form of the mineral periclase. In its purest form, magnesium oxide is a colorless or white crystalline material or very fine powder, with no odor and a bitter taste.

HOW IT IS MADE

A number of methods are available for the preparation of magnesium oxide. Most methods begin with either magnesium hydroxide [Mg(OH)$_2$] or magnesium carbonate (MgCO$_3$), either of which is heated under controlled conditions to produce the desired product. The primary methods of preparation are:

- Dead-burning, in which the hydroxide or carbonate is heated to temperatures ranging from 1,500°C to

Magnesium oxide. Black atom is magnesium; red atom is oxygen; the stick represents a double bond. PUBLISHERS RESOURCE GROUP

2,000°C (3,000° to 4,000°F), so-called because the final product is largely chemically unreactive;

- Hard-burning (also known as caustic-burned), in which the hydroxide or carbonate is heated to somewhat lower temperatures, between 1,000°C and 1,500°C (2,000°F and 3,000°F), allowing the compound to retain some of its chemical reactivity;

- Light-burning, in which the temperature is kept between 700°C and 1,000°C (1,500°F and 2,000°F), resulting in a product with even more reactivity; and

- Fusion, in which the hydroxide or carbonate is heated to temperatures in excess of 2,650°C (4,800°F), producing a very dense, inert product.

An especially pure form of magnesium oxide can be made by taking the product from any of the above reactions and making a slurry with water. A slurry is a mud-like mixture of a solid and liquid that normally do not form a solution. Various chemicals can then be added to the magnesium oxide slurry to remove any contaminants, and the purified slurry is allowed to dry.

COMMON USES AND POTENTIAL HAZARDS

Each form of magnesium oxide has its specialized uses:

- Dead-burned: As refractory brick for cement kilns, furnaces, crucibles, and equipment used in the manufacture of steel;

- Hard-burned: In the production of fertilizers and animal feed, in the extraction of uranium oxide from uranium

Interesting Facts

- Magnesium oxide is sometimes used as a gemstone called periclase. It is found in a wide range of colors from colorless or white to yellow to brown. Its use as a gemstone is somewhat limited, however, because it is not very hard.

ore, as a catalyst, in the manufacture of ceramics, for the tanning of leather, in the synthesis of magnesium compounds, and (its most important single use) in pollution control devices that remove sulfur dioxide from plant exhaust gases;

- Light-burned: In the processing of paper and pulp; as a filler in products made of rubber; and as an ingredient in a host of household and personal care products such as dusting powders, cosmetics, and pharmaceuticals. Some of the best-known pharmaceuticals containing magnesium oxide are antacids used to treat heartburn, upset stomach, or acid indigestion and laxatives for the treatment of constipation or in preparation for surgery.

- Fusion: As refractory linings for electric arc furnaces and in insulating materials used in many household electrical products.

Words to Know

CATALYST A material that increases the rate of a chemical reaction without undergoing any change in its own chemical structure.

REFRACTORY MATERIAL A material with a high melting point, resistant to melting, often used to line the interior of industrial furnaces.

SYNTHESIS A chemical reaction in which some desired chemical product is made from simple beginning chemicals, or reactants.

Contact with magnesium oxide dust or fumes may irritate the skin, eyes, or respiratory system. Symptoms of exposure may include fatigue and lethargy. The greatest concern for such health hazards rests with people who come into contact with the pure product in their line of work.

FOR FURTHER INFORMATION

"Everything You Wanted to Know about Magnesium Oxide." Martin Marietta Magnesia Specialties.
http://www.magspecialties.com/students.htm (accessed on October 14, 2005).

"Magnesia—Magnesium Oxide (MgO)." Azom.com.
http://www.azom.com/details.asp?ArticleID=54 (accessed on October 14, 2005).

See Also Magnesium Hydroxide

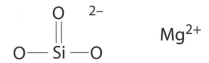

OTHER NAMES:
See Overview.

FORMULA:
$Mg_3Si_4O_{10}(OH)_2$

ELEMENTS:
Magnesium, silicon,
oxygen, hydrogen

COMPOUND TYPE:
Hydrated salt
(inorganic)

STATE:
Solid

MOLECULAR WEIGHT:
379.27 g/mol

MELTING POINT:
1500°C (2700°F);
begins to lose water
of hydration above
900°C (1600°F)

BOILING POINT:
Not applicable

SOLUBILITY:
Insoluble in water and
most organic
solvents

KEY FACTS

Magnesium Silicate Hydroxide

OVERVIEW

Magnesium silicate hydroxide (mag-NEE-zee-um SILL-uh-kate hye-DROK-side) is also known as hydrated magnesium silicate, hydrous magnesium silicate, magnesium silicate hydrous, talc, talcum, and soapstone. It belongs to a large family of magnesium silicates that occur in nature. Magnesium silicates contain at least one magnesium ion and one or more silicate (SiO_3) ions, and often contain one or more molecules of water of hydration. Other members of the family include magnesium metasilicate ($MgSiO_3$), magnesium orthosilicate (Mg_2SiO_4), magnesium trisilicate ($Mg_2Si_3O_8$), and magnesium trisilicate pentahydrate ($Mg_2Si_3O_8 \cdot 5H_2O$).

The naturally-occurring form of magnesium silicate hydroxide is called talc, a soft mineral that feels waxy and soapy to the touch. This characteristic has led to another name for the mineral: soapstone. Talc's chemical formula differs somewhat from that of magnesium silicate hydroxide: $Mg_3Si_4O_{10}(OH)_2$.

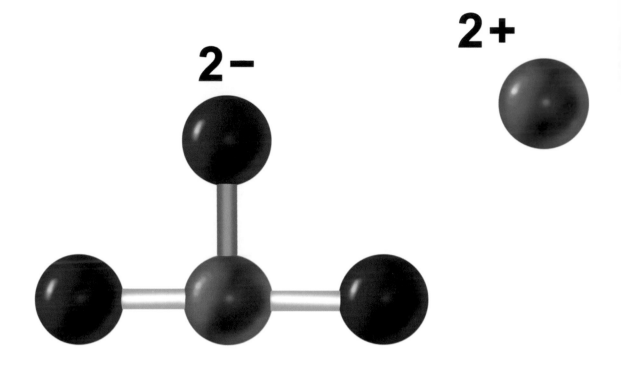

Magnesium silicate. Red atoms are oxygen; turquoise atom is magnesium; and orange atom is silicon. Gray stick indicates a double bond.

Talc is one of the softest minerals known. It has a numerical rank of 1 on the Mohs scale of minerals. The Mohs scale ranks minerals from the softest (1 = talc) to the hardest (10 = diamond). Talc is so soft that it can be scratched with the fingernail. The mineral has a pearly luster and may come in a variety of colors, ranging from white gray, or silver to black, brown, pink, or green. Color variations depend on impurities in the mineral.

Magnesium silicate hydroxide normally occurs as a fine white powder with no odor, insoluble in most solvents, noncombustible, resistant to heat, and with a tendency to absorb moisture from the air.

HOW IT IS MADE

The magnesium silicates are obtained from natural sources, such as talc (magnesium silicate hydroxide), enstatite (magnesium metasilicate), forsterite (magnesium orthosilicate), meerschaum (magnesium trisilicate), or serpentine ($Mg_3Si_2O_7$). In all cases, extraction of the desired compound

Interesting Facts

- The magnesium silicates contain three of the eight most abundant elements in Earth's crust: oxygen (the most abundant element), silicon (the second most abundant element), and magnesium (the eighth most abundant element).

involves a number of steps in which the mineral is separated from waste products, crushed, purified, and processed into its desired form (powder or crystal, for example).

COMMON USES AND POTENTIAL HAZARDS

The form of magnesium silicate hydroxide with which most people are familiar is talc, the main ingredient in talcum powder. Talcum powder is used directly as a skin treatment, especially for babies, primarily as a moisture absorbent and for soothing the skin. It is also used in a number of cosmetic products, including blushes, eye shadows, make-up foundations, and face powders. Talc is also used widely as an additive to give products a smoother texture. Some products in which it is found include paints, rubber products, roofing materials, ceramics and insecticides. Other applications of talc include:

- As a food additive to prevent foods from clumping and sticking together;

- In the polishing of rice;

- In the production of olive oil to improve the product's clarity;

- For countertops in chemical laboratories;

- As a lubricant between sheeted products, such as particleboard, to prevent individual sheets from sticking to each other;

- As a filler in a number of products, including soap, putty, plaster, oilcloth, and rubber products; and

- As a non-caking agent in animal feeds and fertilizers.

Words to Know

ION An atom or a group of atoms with either a positive or negative charge, because it has either lost or gained an electron.

MOHS SCALE A numerical scale used to compare the hardness of a material. Talc has a Mohs value of I; diamond has a value of IO.

REFRACTORY MATERIAL A material with a high melting point, resistant to melting, often used to line the interior of industrial furnaces.

WATER OF HYDRATION Water that has combined with a compound by some physical means.

Magnesium silicate hydroxide and other forms of magnesium silicates have a number of industrial applications, including:

- In the manufacture of glass and ceramic materials;
- As an insulating material for electrical devices;
- As a refractory material in industrial furnaces;
- In clean-up operations following oil spills; and
- As an odor absorbent.

Exposure to talc dust may cause irritation of the skin, respiratory system, and, especially, the eyes. Harmful effects may include skin rash, eye damage, and coughing and wheezing. These symptoms generally occur only as the result of long-term and consistent exposure to dust. There is no evidence that talc is carcinogenic unless contaminated with other minerals, such as asbestos and/or silica. The amount of talc and other magnesium silicates to which the average person is exposed is probably too low to produce any detectable health effects.

FOR FURTHER INFORMATION

"Chemical Summary: Talc." CHEC's HealtheHouse. http://www.checnet.org/healthehouse/chemicals/chemicals-detail.asp?Main_ID=2 (accessed on October 26, 2005).

"Magnesium Silicates." In Pradyot Patnaik. *Handbook of Inorganic Chemicals.* New York: McGraw-Hil, 2003, 534-535.

"Talc." New Jersey Department of Health and Senior Services. http://www.state.nj.us/health/eoh/rtkweb/1773.pdf (accessed on October 26, 2005).

"Talc Mineral Data." Mineral of the Month Club. http://webmineral.com/data/Talc.shtml (accessed on October 26, 2005).

$$O \quad O^-$$
$$\underset{O^- \quad O}{\overset{O \quad O^-}{S}} \qquad Mg^{++}$$

OTHER NAMES:
None

FORMULA:
MgSO₄

ELEMENTS:
Magnesium, sulfur,
oxygen

COMPOUND TYPE:
Inorganic salt

STATE:
Solid

MOLECULAR WEIGHT:
120.40 g/mol

MELTING POINT:
1,127°C (2,061°F);
decomposes

BOILING POINT:
Not applicable

SOLUBILITY:
Soluble in water,
ethyl alcohol, and
glycerol; slightly
soluble in ether

KEY FACTS

Magnesium Sulfate

OVERVIEW

Magnesium sulfate (mag-NEE-zee-um SUL-fate) occurs as the anhydrous salt and in a number of hydrated forms, including $MgSO_4 \cdot H_2O$, $MgSO_4 \cdot 4H_2O$, $MgSO_4 \cdot 5H_2O$, $MgSO_4 \cdot 6H_2O$, and $MgSO_4 \cdot 7H_2O$. The heptahydrate ($MgSO_4 \cdot 7H_2O$) is commonly known as Epsom salts. All of the hydrates occur in nature as the minerals, respectively, kieserite, starkeyite, pentahydrite, hexahydrite, and epsomite. Magnesium sulfate in all forms is a colorless or white crystalline or powdery material with no odor but a bitter taste. The hydrates all lose their water of hydration when heated. For example, the heptahydrate loses one molecule of water spontaneously at room temperature, four molecules of water when heated above 70°C (160°F), six molecules of water at 150°C (300°F), and all seven molecules of water above 200°C (400°F).

Magnesium sulfate heptahydrate has been known and used since the late seventeenth century. The term *Epsom salts* was introduced in 1695 by the English naturalist Nehemiah Grew (1628-1711), who named the compound after

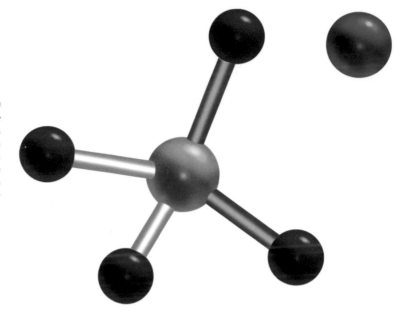

Magnesium sulfate heptahydrate. Red atoms are oxygen; turquoise atom is magnesium; and yellow atom is sulfur. Gray sticks indicate double bonds. PUBLISHERS RESOURCE GROUP

the spring waters near Epsom, England, from which it was often extracted. Grew received a royal patent for the collection, purification, and sale of the compound.

Although Epsom salts were not given their name until the late seventeenth century, their use as a therapy for many disorders was known at least two centuries earlier. People traveled to the Epsom area, as well as to other locations where the compound occurred in spring waters, just to drink the waters. Among its other benefits, Epsom salts was regarded as a purgative, a substance that helps empty the bowels (intestines).

HOW IT IS MADE

All forms of magnesium sulfate can be obtained from natural sources in a very high degree of purity, which is their primary source for commercial and industrial applications. For example, simply evaporating the water from springs like those around Epsom results in crystalline magnesium heptahydrate of sufficient purity for many uses. Water taken from salt lakes may also contain magnesium sulfate in sufficient concentration to allow its extraction by

Interesting Facts

- Up to 2 percent of sea salt may consist of magnesium sulfate.

a process of dilution and evaporation. Anhydrous magnesium sulfate is obtained from kieserite by dissolving the mineral in water at 90°C (200°F), allowing the solution to cool, and then removing the crystals that form by filtration or centrifuging.

COMMON USES AND POTENTIAL HAZARDS

Epsom salts, the heptahydrate of magnesium sulfate, has long been used for its medicinal benefits. When put into bathwater, it helps reduce inflammation and swelling, and relaxes muscles. People still visit hot springs and other types of medicinal waters where magnesium sulfate is one of the dissolved substances that contributes to one's well-being. The compound is also sold over the counter for use at home for similar purposes.

The various forms of magnesium sulfate also have a number of important commercial and industrial uses, including:

- As an animal feed, to ensure that animals receive the magnesium they require in their diets;

- In fertilizers, with the function of treating soil for magnesium deficiencies;

- In the textile industry, where it is used for weighting and sizing silk, as a mordant, for treating finished cotton fabric, and in fireproofing fabrics;

- In the production of ceramics;

- In the preparation of a specialized form of cement known as oxysulfate cement;

- In electroplating processes;

Words to Know

ANHYDROUS Form of a compound that lacks any water of hydration.

CENTRIFUGING A method of separating the components of a mixture by spinning the mixture at a high rate of speed. Centrifugal force acts differently on each component of the mixture, causing them to separate from each other.

ELECTROPLATING A process by which a thin layer of one metal is deposited on top of a second metal by passing an electric current through a solution of the first metal.

HYDRATE A chemical compound formed when one or more molecules of water is added physically to the molecule of some other substance.

MORDANT A substance used in dyeing and printing that reacts chemically with both a dye and the material being dyed to help hold the dye permanently to the material.

SYNTHESIS A chemical reaction in which some desired chemical product is made from simple beginning chemicals, or reactants.

WATER OF HYDRATION Water that has combined with a compound by some physical means.

- As a catalyst for some chemical and industrial operations; and

- For the synthesis of a number of magnesium compounds, such as magnesium stearate, used in specialized types of soaps.

In dry form, magnesium sulfate can cause irritation of the skin, eyes, and respiratory system, producing symptoms such as sneezing and dryness of the mucous membranes (the soft tissues lining the breathing and digestive passages); pain, redness and tearing of the eyes; and nausea, vomiting, abdominal cramps, and diarrhea as a result of ingesting the compound. People who use the compound should use reasonable precaution in handling the material, taking care not to inhale or ingest it.

FOR FURTHER INFORMATION

"About Epsom Salt." Epsom Salt Council.
http://www.epsomsaltcouncil.org/about_epsom_salt.htm
(accessed on October 14, 2005).

"Magnesium Sulfate Anhydrous." J. T. Baker. http://www.jtbaker.com/msds/englishhtml/m0235.htm (accessed on October 14, 2005).

Nardozzi, Charlie. "Fertilize with Epsom Salts." Do It Yourself.com. http://doityourself.com/fertilizer/fertilizeepsomsalts.htm (accessed on October 14, 2005).

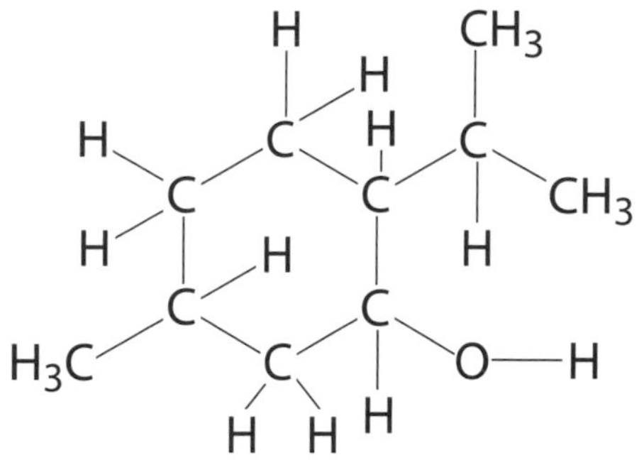

OTHER NAMES:
Hexahydrothymol;
methylhydroxyiso-
propylcyclohexane;
peppermint camphol

FORMULA:
$CH_3C_6H_9 (C_3H_7)OH$

ELEMENTS:
Carbon, hydrogen,
oxygen

COMPOUND TYPE:
Organic

STATE:
Solid

MOLECULAR WEIGHT:
156.26 g/mol

MELTING POINT:
41°C to 43°C (106°F to
109°F)

BOILING POINT:
212°C (414°F)

SOLUBILITY:
Slightly soluble in
water; very soluble in
alcohol, chloroform,
and ether

KEY FACTS

Menthol

OVERVIEW

Menthol (MEN-thol) occurs naturally in the peppermint plant. In pure form it occurs as a white crystalline material with a cooling taste and odor. Peppermint is one of the oldest known herbal remedies. Dried peppermint leaves have been found in Egyptian pyramids dating to at least 1000 BCE, and its use among the Greeks and Romans in cooking and medical preparations is well known. Peppermint was not introduced to western Europe, however, until the eighteenth century, when it was used to treat a variety of ailments ranging from toothaches to morning sickness. It was first brought to the United States about a century later.

HOW IT IS MADE

Peppermint oil is extracted from the leaves of the peppermint plant, *Mentha piperita*, by steam distillation, by which various oils in the plant are separated from each other.

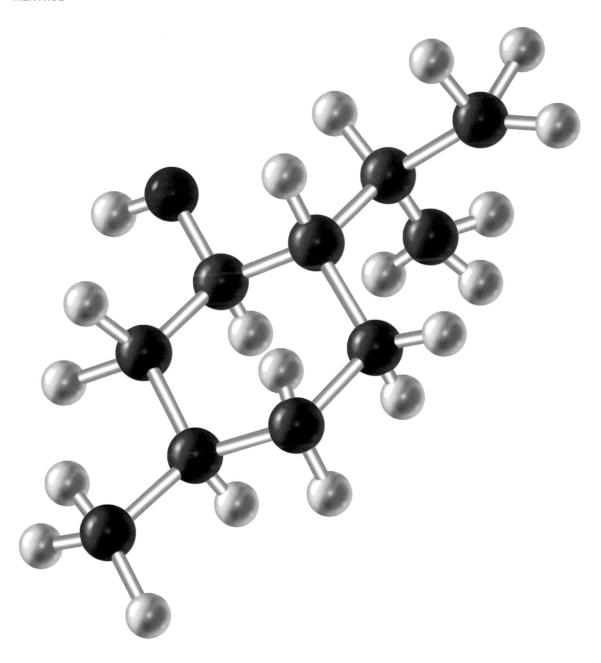

Menthol. Red atom is oxygen; white atoms are hydrogen; and black atoms are carbon.

The peppermint oil is then frozen to extract the menthol from other components of the oil. Menthol can also be produced synthetically by the reduction of thymol [$(CH_3)_2CHC_6H_3(CH_3)OH$] with hydrogen.

Interesting Facts

- Menthol was first added to cigarettes in the early 1900s to make a cooler tasting product. Tobacco companies claimed that menthol cigarettes were less irritating to the throat and an effective way of treating sore throats. Today, about 25 percent of all cigarettes sold in the United States are mentholated.

COMMON USES AND POTENTIAL HAZARDS

Menthol smells like mint and creates a soothing and sometimes tingling sensation when it touches the skin. Scientists theorize that menthol creates the cooling sensation by triggering the same receptors on skin that tell the body's nerves to respond to cold temperatures.

The cooling sensation makes menthol a desirable additive to aftershave lotions, skin cleansers, lotions, sore throat lozenges, and lip balms. Menthol is also used in a variety of cosmetics applied to the skin and medications for the relief of itching. It is also added to foods such as chewing gums and candies to impart a mint-like flavor.

When inhaled or ingested as a lozenge, menthol can relieve nasal congestion and coughs, as well as cool and numb the throat to ease the pain of sore throats. It can also be used in ointments with camphor and eucalyptus to produce cooling and antiseptic properties. These ointments can be applied to the chest and/or nostrils to clear the nose and reduce coughing. One of the most famous menthol-containing products is Vicks® VapoRub, which is used to relieve coughs and congestion.

Although menthol is soothing and cooling in small quantities, it produces a quite different effect in larger quantities. Gargling with a large amount of menthol-containing mouthwash, for example, can create an unpleasant burning sensation.

Although menthol has been classified as a "generally recognized as safe" (GRAS) product and approved for use in foods by the U.S. Good and Drug Administration, some side effects have been reported. On contact with the skin,

Words to Know

DISTILLATION A process of separating two or more substances by boiling the mixture of which they are composed and condensing the vapors produced at different temperatures.

REDUCTION The chemical reaction in which oxygen is removed from

a substance or electrons are added to a substance.

SYNTHESIS A chemical reaction in which some desired chemical product is made from simple beginning chemicals, or reactants.

menthol may cause irritation. Ingesting large quantities can cause abdominal pain, nausea, vomiting, dizziness, drowsiness, and even coma. These effects are more likely to occur in infants and children than in adults.

FOR FURTHER INFORMATION

"Cool Menthol 1" and "Cool Menthol 2." Great Moments in Science. http://www.abc.net.au/science/k2/moments/s537539.htm and http://www.abc.net.au/science/k2/moments/s537548.htm (accessed on October 12, 2005).

Cummings, Linda Scott. "M2258 Menthol." Paleoresearch Institute. http://www.paleoresearch.com/MSDS/MENTHOL%20-%20 SIGMA%20CHEMICAL%20-%2005-19-1997.htm (accessed on October 12, 2005).

"Menthol and Tobacco Smoking." http://goodhealth.freeservers.com/MethTobaccoIntro.html (accessed on October 12, 2005).

"Menthol—Its Chemistry and Many Uses." http://goodhealth.freeservers.com/MenthUseThisOne.htm (accessed on October 12, 2005).

Travis, J. "Cool Discovery: Menthol Triggers Cold-Sensing Protein." *Science News* (February 16, 2002): 101-102.

See Also Camphor

$$S = Hg$$

OTHER NAMES:
Mercuric sulfide;
cinnabar; vermillion;
Chinese red

FORMULA:
HgS

ELEMENTS:
Mercury, sulfur

COMPOUND TYPE:
Inorganic salt

STATE:
Solid

MOLECULAR WEIGHT:
232.66 g/mol

MELTING POINT:
Data differ
significantly; see
Overview

BOILING POINT:
Not applicable

SOLUBILITY:
Insoluble in water,
alcohol, and most
acids; soluble in
aqua regia

K E Y   F A C T S

Mercury(II) Sulfide

OVERVIEW

Mercury(II) sulfide (MER-kyuh-ree two SUL-fide) occurs in two forms, red and black. Red mercury(II) sulfide, commonly known as cinnabar, begins to change color when heated to temperatures of about 250°C (500°F) and converts to the black form at 386°C (727°F). If heated further, it sublimes (changes directly from a solid to a gas without first melting) at 583.5°C (1,082°F). If allowed to cool, it then reverts to its original reddish color. Black mercury(II) sulfide goes through a similar process, changing to its red counterpart before melting at 583.5°C (1,082°F). Some authoritative resources give significantly different temperatures for these transitions. Red mercury(II) sulfide occurs naturally as the mineral cinnabar, while black mercury(II) sulfide occurs only rarely in nature, then as the mineral metacinnabar (meaning "similar to cinnabar").

HOW IT IS MADE

Red mercury(II) sulfide is obtained commercially from the mineral cinnabar. The compound can also be made

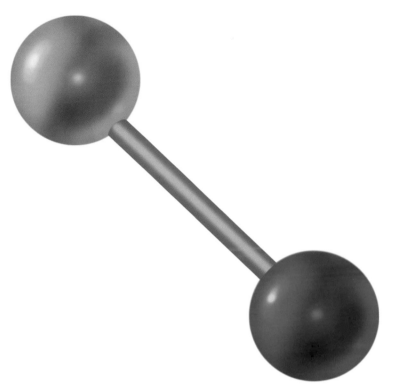

Mercury Sulfide. Turquoise atom is mercury and yellow atom is sulfur. Stick is a double bond. PUBLISHERS RESOURCE GROUP

synthetically by heating mercury and sulfur together in a gaseous state or by heating mercury with a solution of potassium pentasulfide (K_2S_5). The compound produced by either of these methods is commonly known as English vermillion, or simply, vermillion. The term vermillion is generally reserved for any form of mercury(II) sulfide that has been made synthetically rather than extracted from cinnabar. Other methods for the preparation of both red and black mercury(II) sulfide are available. For example, the black form can be produced by reacting sodium thiosulfate ($Na_2S_2O_3$) with sodium mercurichloride (Na_2HgCl_4).

COMMON USES AND POTENTIAL HAZARDS

The earliest records of the use of mercury(II) sulfide by humans date to about the third millennium BCE in China, where the compound was used to cure diseases, relieve pain, as a narcotic and an antiseptic, and as a preservative. Chinese alchemists referred to the compound as "celestial

Interesting Facts

- Women used vermillion during the Renaissance period to redden their lips and cheeks.

- Mercury(II) sulfide is a comparatively expensive compound, selling in late 2005 for about 1,600 dollars per 100 grams.

sands" or "god's sand" and believed that it could transform base metals, like iron and lead, into precious metals, like silver and gold.

Today, the primary use of mercury(II) sulfide is in the production of metallic mercury. The sulfide is heated in a furnace to temperatures of 600°C to 700°C (1,100°F to 1,300°F), resulting in the formation of sulfur dioxide and mercury metal. In a second process, the sulfide is treated with lime (CaO), resulting in the formation of mercury metal, calcium sulfide (CaS) and calcium sulfate ($CaSO_4$).

The other major use for mercury(II) sulfide, in either red or black form, is as a pigment in artists' paints, for coloring paper and plastics, and for marking linen. The black form is also used as a pigment for the coloring of rubber, horn, and other materials. Red mercury(II) sulfide finds some use also as an antibacterial agent.

Both forms of mercury(II) sulfide are highly toxic by ingestion, inhalation, or absorption through the skin. Some of the symptoms resulting from mercury(II) sulfide poisoning include inflammation and itching of the skin; redness,

Words to Know

ALCHEMY An ancient field of study from which the modern science of chemistry evolved.

AQUA REGIA A combination of concentrated nitric acid and hydrochloric acid.

itching, burning, and watering of the eyes; excessive saliva-tion; pain when chewing; gingivitis with loosening of teeth; and mental disturbances, such as loss of memory, insomnia, irritability, and vague fears of depression. Anyone who has been exposed to mercury(II) sulfide and experiences such symptoms requires immediate medical attention.

FOR FURTHER INFORMATION

Liu, Guanghua. "Chinese Cinnabar." *The Mineralogical Record* (January-February 2005): 69-80.

"The Mineral Cinnabar." Amethyst Galleries. http://mineral.galleries.com/minerals/sulfides/cinnabar/cinnabar.htm (accessed on October 14, 2005).

$$\begin{array}{ccc} H & & H \\ & \diagdown \; \diagup & \\ & C & \\ & \diagup \; \diagdown & \\ H & & H \end{array}$$

Methane

OTHER NAMES:
Marsh gas; methyl
hydride

FORMULA:
CH₄

ELEMENTS:
Carbon, hydrogen

COMPOUND TYPE:
Hydrocarbon; alkane
(organic)

STATE:
Gas

MOLECULAR WEIGHT:
16.04 g/mol

MELTING POINT:
−182.47°C
(−296.45°F)

BOILING POINT:
−161.48°C
(−322.63°F)

SOLUBILITY:
Very slightly soluble
in water and acetone;
soluble in ethyl
alcohol, methyl
alcohol, and ether

KEY FACTS

OVERVIEW

Methane (METH-ane) is a colorless, odorless, tasteless, flammable gas that is less dense then air. It is the primary component of natural gas. Methane is the simplest of all hydrocarbons, organic compounds that contain carbon and hydrogen and no other elements.

HOW IT IS MADE

Methane formed millions of years ago from microscopic underwater plants and bacteria that dropped to the bottom of the ocean when they died. Over millions of years, they were crushed and heated by the pressure of layers of sand, dirt, and other materials that accumulated on top of them. The mineral components of the undersea mud gradually turned into a type of rock known as shale. Some of the organic components turned into natural gas, which is mostly methane. The natural gas became trapped in porous rocks called reservoir rocks and in larger pockets of the rock called reservoirs or geologic traps. Natural gas is now found in

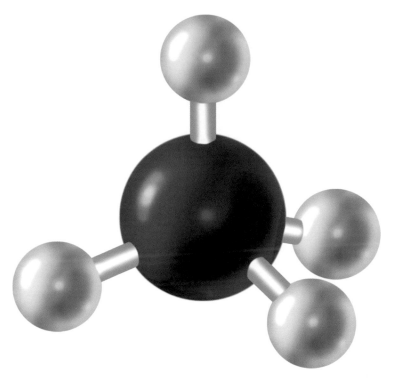

Methane. Black atom is carbon and white atoms are hydrogen. PUBLISHERS RESOURCE GROUP

pockets by itself, but is more commonly found floating on top of petroleum lakes in underground reservoirs.

Methane is also found in conjunction with pockets of coal. The largest reserves of natural gas in the United States are in Texas, Alaska, Oklahoma, Ohio, and Pennsylvania. Oil and gas companies remove natural gas from the ground by drilling. They then purify the natural gas by separating the components of which it is made, such as methane, ethane, propane, and butane. After isolation from natural gas, methane is often liquefied, which makes its easier to store and transport.

Although abundant supplies of methane exist, it can also be produced synthetically. For example, the reaction between steam and hot coal results in the formation of synthesis gas, a mixture of hydrogen and carbon monoxide. When this mixture is passed over a catalyst containing nickel metal, methane is formed. A very similar process, called the Sabatier process, uses a mixture of hydrogen and carbon dioxide, rather than carbon monoxide, also resulting in the formation

Interesting Facts

- Methane is sometimes called marsh gas because it forms in swamps as plants and animals decay under water.

- Methane is odorless, but gas companies add traces of sulfur-containing compounds with strong odors so that people will be able to smell gas leaks and avoid suffocation or explosions.

- Some experts estimate that enough methane is present in the Earth's surface to last as much as two hundred years, although extraction of some methane resources may prove to be difficult.

of methane. Finally, methane produced during the anaerobic decomposition of manure can be captured and purified.

COMMON USES AND POTENTIAL HAZARDS

When methane burns, it releases a large amount of energy, making it useful as a fuel. Humans have known about methane as a source of energy for thousands of years. Temples in the ancient world often burned "eternal flames" that may have been fueled by natural gas. In the early nineteenth century, people began using natural gas as a light source. Once oil was discovered in the 1860s, however, its use, and the electricity produced by burning oil, became much more popular, and people abandoned natural gas as a fuel except for limited use in cooking.

Natural gas has become more popular in recent years because its use results in less pollution than petroleum and other fossil fuels. Some uses include heating homes, offices, and factories; powering room heaters and air conditioners; and operating home appliances such as water heaters and stoves.

Scientists are now exploring other uses for methane and natural gas with the hope that they might eventually become the most important fuels used by humans. Methane has some advantages over petroleum and coal as a fuel. It

Words to Know

ANAEROBIC Describing a process that takes place in the absence of oxygen.

GREENHOUSE GAS One of several gases, including carbon dioxide and ozone, that causes the greenhouse effect on Earth.

burns more cleanly than either of these other fossil fuels, producing only carbon dioxide and water as combustion products. Some experts believe that methane could be used as a power source of fuel cells, cells that burn hydrogen to produce electricity. Adding natural gas to oil- or coal-fired burners would also help reduce the greenhouse gas emissions of these appliances.

In addition to its applications as a fuel, methane is used in the manufacture of a number of organic and inorganic compounds. For example, ammonia, which is the tenth most important chemical compounds in the United States, based on quantity produced, is made from hydrogen and nitrogen gases. Over 90 percent of the hydrogen used to make ammonia is now obtained by reacting methane with water at high temperatures over a catalyst of iron oxide (Fe_3O_4). Other compounds produced from methane include methanol (methyl alcohol), acetylene (ethyne), formaldehyde (methanal), hydrogen cyanide, carbon tetrachloride, chloroform, methylene chloride, and methyl chloride.

Methane is not toxic, but it can cause suffocation by reducing or eliminating the oxygen a person needs to breathe normally. The primary hazard posed by the gas is its flammability and explosive tendency.

FOR FURTHER INFORMATION

"Chemical of the Week: Methane." Science Is Fun. http://scifun.chem.wisc.edu/chemweek/methane/methane.html (accessed on October 17, 2005).

"Methane." U.S. Environmental Protection Agency. http://www.epa.gov/methane/ (accessed on October 17, 2005).

"Methane Madness: A Natural Gas Primer." The Coming Global Oil Crisis.
http://www.oilcrisis.com/gas/primer/ (accessed on October 17, 2005).

Sherman, Josepha, and Steve Brick. *Fossil Fuel Power.* Mankato, MN: Capstone Press, 2003.

See Also Propane

HO———CH$_3$

OTHER NAMES:
Methanol; wood
alcohol; wood spirit;
carbinol

FORMULA:
CH$_3$OH

ELEMENTS:
Carbon, hydrogen,
oxygen

COMPOUND TYPE:
Alcohol (organic)

STATE:
Liquid

MOLECULAR WEIGHT:
32.04 g/mol

MELTING POINT:
−97.53°C (−143.6°F)

BOILING POINT:
64.6°C (148°F)

SOLUBILITY:
Miscible with water,
ethyl alcohol, ether,
acetone, and many
other organic
solvents

KEY FACTS

Methyl Alcohol

OVERVIEW

Methyl alcohol (METH-uhl AL-ko-hol) is a clear, colorless, flammable, toxic liquid with a slightly alcoholic odor and taste. Methyl alcohol is the simplest alcohol, a family of organic compounds characterized by the presence of one or more hydroxyl (-OH) groups.

HOW IT IS MADE

Methyl alcohol occurs naturally in plants and animals, including humans, as the product of metabolic reactions that occur in all organisms. It also occurs in the atmosphere as the result of the decomposition of dead organisms in the soil. None of these sources is utilized for the commercial production of methyl alcohol. Instead, the primary method for the preparation of methyl alcohol is to react carbon monoxide with water at a temperature of about 250°C (480°F) and pressures of 50 to 100 atmospheres over a mixed catalyst of copper, zinc oxide, and aluminum oxide. Efforts are being made to develop other methods of synthesizing methyl alcohol. In one process, for

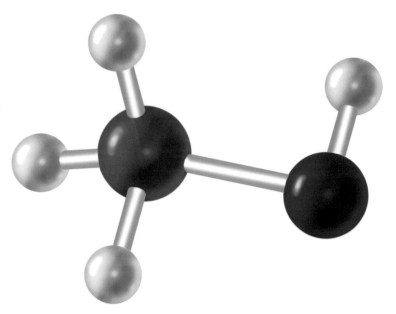

Methyl alcohol. Red atom is oxygen; white atoms are hydrogen; and black atom is carbon. PUBLISHERS RESOURCE GROUP

example, simple hydrocarbons, such as methane, are oxidized over a catalyst of molybdenum metal to produce the alcohol. None of the experimental methods developed for the production of methyl alcohol can yet compete with the traditional carbon monoxide-hydrogen process, however.

COMMON USES AND POTENTIAL HAZARDS

Consumption of methyl alcohol in the United States for 2005 reached about 12 billion liters (3 billion gallons). The largest demand for the compound was in the production of MTBE (methyl-*tert*-butyl ether), a gasoline additive used to improve the efficiency with which a fuel burns and to reduce pollutants released to the atmosphere. Demand for the additive increased rapidly after the U.S. Congress passed the 1990 Clean Air Act Amendments requiring significant reductions in the release of certain pollutants into the atmosphere. A decade later, however, serious questions were being raised about possible serious environment dangers posed by MTBE released into the soil. In the last few years, enthusiasm for the use of MTBE has begun to disappear and a number of states have adopted bans on its use as a gasoline additive. As a result of these actions, the demand for methyl alcohol in producing MTBE has dropped dramatically in the last few years.

Interesting Facts

- At one time, the primary method for making methyl alcohol was to heat wood in a closed space, accounting for the compound's common and popular name of "wood alcohol."

- Methyl alcohol was first isolated, although not in a pure form, by the English chemist and physicist Robert Boyle (1627-1691) although the compound was not synthesized for another two centuries. It was then produced by the French chemist Pierre Eugène Marcelin Berthelot (1827-1907).

The next most important demand for methyl alcohol is as a raw material in the synthesis of many important organic compounds, including formaldehyde; acetic acid; chloromethanes, compounds in which the hydroxyl group and/or one or more hydrogen has been replaced by fluorine, chlorine, bromine, and/or iodine; methyl methacrylate, a compound from which acrylic plastics are made; methylamines, the source of another important class of plastics, dimethyl terephthalate, the monomer for yet another class of plastics; and other products.

Relatively small quantities of methyl alcohol are used in a number of other applications, including:

- As a solvent for household and industrial products;

- As a deicing agent;

- In the preparation of embalming fluids;

- As a softening agent for plastics;

- As a fuel for camp stoves, soldering torches, and race cars;

- In paint removing products;

- As an antifreeze and windshield washing fluid; and

- In the manufacture of a number of pharmaceuticals, including streptomycin, vitamins, and hormones.

Words to Know

METABOLISM A biological process that includes all of the chemical reactions that occur in cells by which fats, carbohydrates, and other compounds are broken down to produce energy and the compounds needed to build new cells and tissues.

MISCIBLE able to be mixed; especially applies to the mixing of one liquid with another.

MONOMER A small molecule used in polymerization reactions to produce very large molecules in which the monomer is repeated hundreds or thousands of times.

SOLVENT A liquid that dissolves another substance.

SYNTHESIS A chemical reaction in which some desired chemical product is made from simple beginning chemicals, or reactants.

Methyl alcohol poses both safety and health hazards. It is highly flammable and, with the appropriate mixture of air, explosive. It is also very toxic by ingestion, producing a variety of effects that include blurred vision, headache, dizziness, drowsiness, and nausea. Although a fatal dose is usually in the range of 100 to 250 mL, cases have been reported in which a person has died after consuming less than 30 mL of the compound. People who work with methyl alcohol in their jobs, including bookbinders, dyers, foundry workers, gilders, hat makers, ink makers, laboratory technicians, painters, photoengravers, and chemical workers, are especially at risk from methanol poisoning. The ready availability of the compound and products in which it is an ingredient means that everyone who uses such products should be aware of the health risks involved in its use. Medical attention is required immediately in case of the ingestion of methyl alcohol.

FOR FURTHER INFORMATION

"Material Safety Data Sheet: Methyl Alcohol, Reagent ACS, 99.8% (GC)." Department of Chemistry, Iowa State University. http://avogadro.chem.iastate.edu/MSDS/methanol.htm (accessed on October 17, 2005).

McGrath, Kimberley A. "Methanol." *World of Scientific Discovery,* 2nd edition. Detroit, MI: Gale, 1999.

"Methanol." U.S. Environmental Protection Agency. Technology Transfer Network, Air Toxics Website. http://www.epa.gov/ttn/atw/hlthef/methanol.html (accessed on October 17, 2005).

Salocks, Charles, and Karlyn Black Kaley. "Methanol." Technical Support Document: Toxicology Clandestine Drug Labs/ Methamphetamine, Volume 1, Number 10. http://www.oehha.ca.gov/public_info/pdf/TSD%20Methanol%20 Meth%20Labs%2010'8'03.pdf (accessed on October 17, 2005).

U. S. Department of Health and Human Services. "Methanol Toxicity." *American Family Physician* (January 1993): 163-171.

See Also Carbon Monoxide; Formaldehyde

$$H_3C \text{---} S \text{---} H$$

OTHER NAMES:
Methanethiol;
mercaptomethane;
thiomethyl alcohol;
methyl sulfhydrate

FORMULA:
CH_3SH

ELEMENTS:
Carbon, hydrogen,
sulfur

COMPOUND TYPE:
Mercaptan; thiol
(organic)

STATE:
Gas

MOLECULAR WEIGHT:
48.11 g/mol

MELTING POINT:
−123°C (−189°F)

BOILING POINT:
5.9°C (43°F)

SOLUBILITY:
Slightly soluble in
water; soluble in
ethyl alcohol and
ether

KEY FACTS

Methyl Mercaptan

OVERVIEW

Methyl mercaptan (METH-uhl mer-KAP-tan) is a colorless, highly flammable, foul smelling gas with the odor of rotten cabbage released from decaying animal and vegetable matter. It is also produced in the intestinal tract by the action of bacteria on a variety of proteins known as the albumins.

Methyl mercaptan belongs to a class of organic compounds called mercaptans or thiols in which one or more sulfhydryl (-SH) groups are attached to a carbon atom. Methyl mercaptan has only one carbon atom, but some mercaptans contain up to twenty carbon atoms. Like methyl mercaptan, other mercaptan compounds are known for their disagreeable odors. For example, allyl mercaptan has the characteristic smell of garlic, while butyl mercaptan occurs in the spray that skunks release to protect themselves from predators.

HOW IT IS MADE

Methyl mercaptan is made by the direct reaction between methanol (methyl alcohol; CH_3OH) and hydrogen sulfide gas

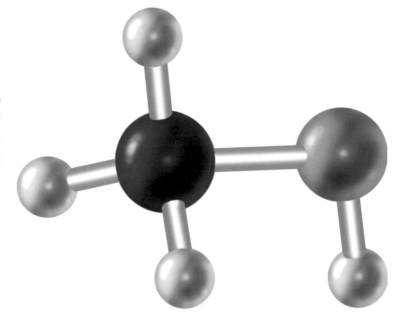

Methyl mercaptan. White atoms are hydrogen; black atom is carbon; and yellow atom is sulfur. PUBLISHERS RESOURCE GROUP

(H_2S). The hydroxyl group (-OH) from the alcohol combines with one hydrogen from hydrogen sulfide to form water, leaving the methyl mercaptan behind as the major product.

COMMON USES AND POTENTIAL HAZARDS

Methyl mercaptan's primary use is in the synthesis of other organic compounds, especially pesticides, fungicides, jet fuel components, plastics, and the amino acid methionine [$CH_3SCH_2CH_2CG(NH_2)COOH$]. Amino acids are the building-block compounds from which proteins are made. The compound is also used as an odorant, a substance with a noticeable and usually offensive odor added to odorless compounds for the purpose of safety. For example, propane and natural gas are two widely used gases that are very flammable, but odorless. If these gases were used without having an odorant added, consumers might not be aware of a leak until the gas caught fire or exploded. The presence of the odorant, such as methyl mercaptan, makes a leak obvious and allows it to be repaired before an accident occurs.

In spite of the fact that it is a natural product, methyl mercaptan is a health and environmental risk. When inhaled, it can irritate the eyes, skin, nose, throat, and lungs, producing

Interesting Facts

- Methyl mercaptan may be responsible for some of the unpleasant odors produced by humans, such as bad breath, flatulence, and stinky feet. When bacteria attack proteins in the body, they release several gases, methyl mercaptan among them. These gases are responsible for bad breath and periodontal disease (inflammation of the gums) in the mouth, and unpleasant odors from other parts of the body. For example, old or dirty socks and shoes are ideal breeding sites for the bacteria responsible for the production of methyl mercaptan and similar foul-smelling organic compounds. One popular product designed to deal with this problem is shoe insoles that contain activated charcoal. Activated charcoal is a form of charcoal consisting of very fine grains that absorbs large volumes of gases, such as methyl mercaptan and its bad-smelling cousins.

- Why does eating asparagus produce bad-smelling urine? Asparagus contains the amino acid methionine, which is metabolized in the body to produce methyl mercaptan. The peculiar odor of urine produced after consuming asparagus is caused by the methyl mercaptan excreted from the body in urine.

dizziness, headache, vomiting, muscle weakness, and loss of coordination. In large concentrations, methyl mercaptan can damage the central nervous system, leading to respiratory failure and even death. The U.S. Occupational Safety and Health Administration (OSHA) has set maximum exposure limits at 20 milligrams of the compound per cubic meter of air per eight-hour work day.

FOR FURTHER INFORMATION

"Bad Breath: What Usually Is the Source of a Person's Bad Breath?" Animated-Teeth.com.
http://www.animated-teeth.com/bad_breath/t3_causes_of_halitosis.htm (accessed on October 17, 2005).

Emsley, John. "Molecules at an Exhibition." http://www.nytimes.com/books/first/e/emsley-molecules.html (accessed on October 17, 2005).

"Safety Data Sheet: Methyl Mercaptan." Air Liquide. http://www.airliquide.com/safety/msds/en/083_AL_EN.pdf (accessed on October 17, 2005).

"ToxFAQs for Methyl Mercaptan." Agency for Toxic Substances and Disease Registry. http://www.atsdr.cdc.gov/tfacts139.html (accessed on October 17, 2005).

$$H_3C\text{—}O\text{—}C(CH_3)_3$$

OTHER NAMES:
MTBE; see Overview
for additional names

FORMULA:
$(CH_3)_3COCH_3$

ELEMENTS:
Carbon, hydrogen,
oxygen

COMPOUND TYPE:
Ether

STATE:
Liquid

MOLECULAR WEIGHT:
88.15 g/mol

MELTING POINT:
$-108.6°C$ ($-163.5°F$)

BOILING POINT:
$55.0°C$ ($131°F$)

SOLUBILITY:
Soluble in water; very
soluble in ethyl
alcohol and ether

Methyl-t-butyl Ether

OVERVIEW

Methyl-t-butyl ether (METH-el TER she-air-ee BYOO-till EE-thur) is a volatile (evaporates easily), colorless, flammable liquid that forms an azeotropic mixture with water. Azeotropic mixtures are combinations of two or more liquids that boil at the same temperature and, therefore, cannot be easily separated from each other.

MTBE was first synthesized in the 1960s by researchers at the Atlantic Richfield Corporation (now ARCO) as an additive designed to increase the octane number (fuel efficiency) of gasoline. The compound was created as a replacement for tetraethyl lead ($Pb(C_2H_5)_4$), which had long been added to gasolines to improve their octane number. Tetraethyl lead was commonly called simply "lead." From 1973 until 1996, lead was gradually removed from most gasoline because of its dangerous environmental and health effects. In 1990, a new use for MTBE was found. In that year, legislation passed by the U.S. Congress required that changes be made in the composition of gasoline so that it would burn more cleanly

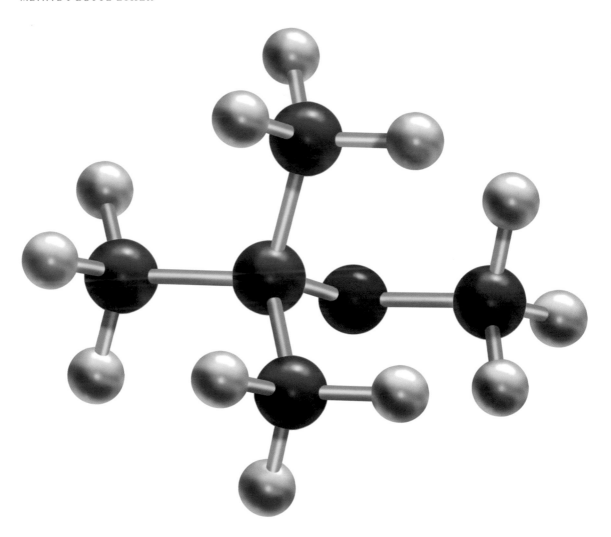

Methyl-t-butyl Ether (MTBE). Black atoms are carbon; red atom is oxygen; white atoms are hydrogen. All sticks are single bonds. PUBLISHERS RESOURCE GROUP

and release fewer pollutants into the atmosphere. One way for companies to meet this regulation was to add oxygenates to their gasoline. Oxygenates are chemical compounds that contain oxygen. They react with fuels, giving off their oxygen and increasing the efficiency with which the fuels burn.

MTBE rapidly became a very popular chemical in the automotive fuel industry. Production rose at a rate of about 7 percent per year in the 1980s. It increased from about 500 million kilograms (1 billion pounds) in 1983 to about 3 billion kilograms (6 billion pounds) in 1990. In that year, it ranked twenty-fourth among all chemicals produced in the United States.

Interesting Facts

- Virtually all of the MTBE produced is used as a fuel additive. If proposed bans and limitations go into effect, there will no longer be any demand for the chemical. As such, production is likely to drop nearly to zero. One company that surveys chemical production predicts that the production of MTBE is likely to decrease at the rate of about 8 percent per year over the foreseeable future.

In the early 1990s, however, MTBE began showing up in groundwater in a number of states. Upon investigation, researchers discovered that MTBE had leaked out of underground tankers, had been spilled during transportation, and had escaped into the environment in other ways. It sank into the ground, where it mixed with groundwater forming azeotropic mixtures that could not be easily separated. The compound was found in wells, lakes, streams, and public water supplies.

These findings raised concerns because some experts believe that MTBE is a health hazard to humans and other animals. Although strong evidence is not yet available, many authorities believe that MTBE is a carcinogen and that it may be responsible for other health problems and damage to the environment. These concerns have led to legislative and administrative action banning or limiting the use of MTBE in gasoline. The first such action occurred in 1999 when Governor Gray Davis of California announced a program to cut back and eventually eliminate the use of MTBE in the state. A year later, the U.S. Environmental Protection Agency announced similar plans for the nation as a whole.

MTBE is also known by the following names: Methyl-*tert*-butyl ether; t-butyl-methyl ether; and *tert*-buty-methyl ether.

HOW IT IS MADE

MTBE is made by reacting methanol (methyl alcohol; CH_3OH) with isobutylene (isobutene; $CH_3C(CH_3)=CH_2$). The

Words to Know

AZEOTROPIC MIXTURE A combination of two or more liquids that boil at the same temperature and, therefore, cannot be easily separated from each other.

CARCINOGEN A chemical that causes cancer in humans or other animals.

OXYGENATES Chemical compounds that contain oxygen.

SOLVENT A substance that is able to dissolve one or more other substances.

VOLATILE Able to turn to vapor easily at a relatively low temperature.

formula for this reaction is $CH_3OH + CH_3C(CH_3)=CH_2 \rightarrow (CH_3)_3COCH_3$.

COMMON USES AND POTENTIAL HAZARDS

More than 99 percent of the MTBE produced is used as a gasoline additive. The remaining quantity is used as a solvent and in the production of other chemical compounds.

MTBE is a mild irritant to the skin, eyes, and respiratory system. In larger doses, it may cause more serious problems, including nausea, vomiting, diarrhea, gastrointestinal problems, headache, dizziness, and loss of balance and coordination. In extreme cases, MTBE may produce severe lung damage, respiratory failure, convulsions, respiratory arrest (breathing stops), unconsciousness, coma, and death. The compound has been found to cause cancer in laboratory animals. No similar studies are available for humans, but experts tend to agree that MTBE is likely carcinogenic in humans also.

FOR FURTHER INFORMATION

"Groundwater Protection." American Petroleum Institute. http://api-ep.api.org/environment/index.cfm?objectid=9EC44CD5-E167-49C4-8EBE7B354E4B3CD9&method=display_body&er=1&bitmask=00200800800000000000 (accessed on December 29, 2005).

"Methyl tert-Butyl Ether (MTBE) and Other Gasoline Oxygenates." U.S. Geological Survey. http://sd.water.usgs.gov/nawqa/vocns/mtbe.html (accessed on December 29, 2005).

"Methyl Tertiary Butyl Ether (MTBE)." U.S. Environmental Protection Agency. http://www.epa.gov/mtbe/ (accessed on December 29, 2005).

"MTBE (methyl tert-butyl ether)." The Innovation Group. http://www.the-innovation-group.com/welcome.htm (accessed on December 29, 2005).

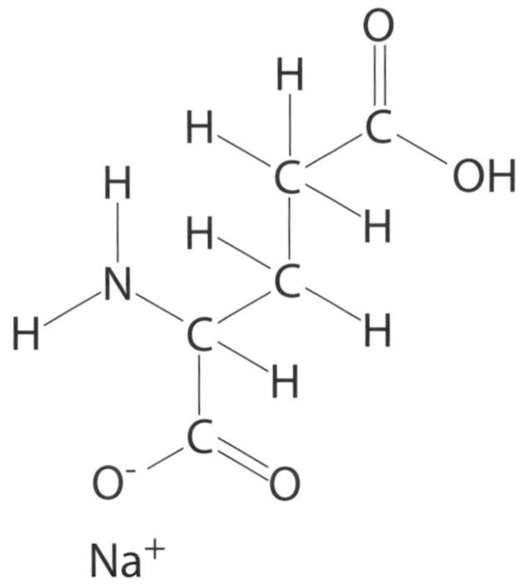

Na$^+$

KEY FACTS

Monosodium Glutamate

OVERVIEW

Monosodium glutamate (mon-oh-SOH-dee-yum GLOO-tuh-mate) is an almost completely odorless white crystalline powder. It is the sodium salt of a common amino acid called glutamic acid. An organic salt is a compound formed when an inorganic base, such as sodium hydroxide, reacts with an organic acid, such as glutamic acid.

Monosodium glutamate has been available as a commercial product for about a century. But the compound has been used in its natural form for much longer. The ancient Greeks and Romans used fish sauce, which contains glutamic acid as a natural ingredient, in their cooking. Later Europeans also used a form of the substance in a product known as garum.

The German chemist Karl Heinrich Ritthausen (1826-1912) first identified glutamic acid in wheat gluten in 1866, and its chemical structure was first identified in 1890 by the German chemist Wolff. Then in 1908, the Japanese chemist Kikunae Ikeda (1864-1936) discovered the flavor-enhancing properties of glutamic acid. He found in seaweed broth a

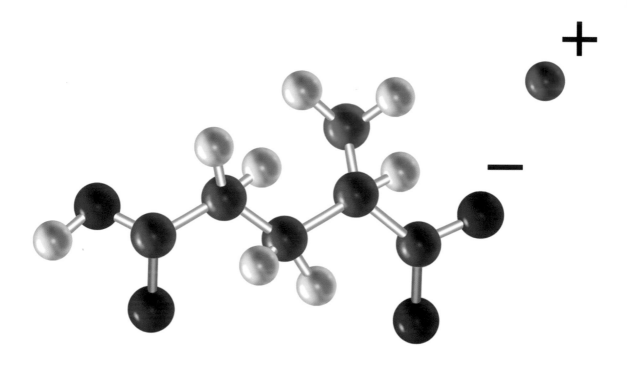

Monosodium glutamate. Red atoms are oxygen; white atoms are hydrogen; black atoms are carbon; blue atom is nitrogen; and turquoise atom is sodium. Gray sticks indicate double bonds. PUBLISHERS RESOURCE GROUP

compound, later identified as glutamic acid, responsible for the unique taste of cheese, meat, and tomatoes. The compound was not sweet, salty, sour, or bitter, but instead had a rich meaty taste. He named the unusual flavor *umami*. Ikeda used glutamic acid crystals to make a seasoning, which he then patented. Like salt and sugar, it was readily soluble in water and could be stored for long periods of time without clumping. The seasoning was actually the sodium salt of glutamic acid, monosodium glutamate, first sold commercially in Japan under the name *Ajinomoto*, which means "essence of taste." The product was introduced into the United States shortly after World War II, and the U.S. Food and Drug Administration approved its use as a food additive in 1958.

HOW IT IS MADE

Glutamic acid in the form of glutamate exists in two forms: bound and free. The term bound glutamate refers to the glutamic acid that has combined with other amino acids to form proteins. The term free glutamate refers to the acid radical $COOH(CH_2)_2CH(NH_2)COO^-$ formed when a single molecule of glutamic acid loses one hydrogen ion. Foods such as

Interesting Facts

- The flavor-enhancing qualities of monosodium glutamate are a function of its concentration in foods. At concentrations of less than about 0.5 percent, it enhances the flavor of a food product. At greater concentrations, it begins to make the food product taste bad.

tomatoes, mushrooms, and cheeses are naturally high in glutamate, which is responsible for their strong flavors. Only the free form of glutamate enhances food flavors. In the human body, glutamate is a nonessential amino acid found in the brain, muscles, kidneys, liver, and other organs.

Monosodium glutamate is made commercially by the hydrolysis of the waste products of beet sugar refining, wheat, or corn gluten. Hydrolysis is the process by which a compound reacts with water to form two new compounds. In the preparation of monosodium glutamate, water breaks apart the proteins present in the raw materials used (such as wheat or corn gluten), freeing the amino acids of which they are made. The glutamic acid present in the proteins can then be separated from other amino acids present, then isolated and purified. Monosodium glutamate can also be made by a very similar process in which bacterial fermentation is used to break proteins down into their component parts, of which glutamic acid is one, a process which is now the primary means of production for the compound.

COMMON USES AND POTENTIAL HAZARDS

The sole use of monosodium glutamate is as a flavor enhancer in food products. Annual worldwide consumption of the additive has been estimate at about 1 million metric tons (1.1 million short tons). It is used primarily in a variety of Asian foods, including soups, canned foods, and processed meats.

Monosodium glutamate is considered safe for human consumption. It is on the U.S. Food and Drug Administration's Generally Recognized as Safe (GRAS) list. The GRAS list contains chemicals that have never been tested scientifically for safety, but are generally believed to be safe for human consumption. In 1987, health experts from the United Nations and the World Health Organization reviewed more than two hundred studies on MSG and determined that the compound is a safe food additive when used at customary levels.

In spite of these studies and rulings, many people believe that monosodium glutamate can have harmful health effects. They say that eating foods that include MSG can cause headache, nausea, sweating, rapid heartbeat, a burning sensation in the back of the neck, and/or tightness in the chest. Thus far, scientific studies have been unable to confirm the association of MSG with these symptoms.

Researchers theorize that the reactions may be the result of allergies to certain ingredients in the foods. In particular, they are likely to occur among people who have specific allergies to monosodium glutamate or who have asthma.

FOR FURTHER INFORMATION

"The Facts on Monosodium Glutamate." European Food Information Council.
http://www.eufic.org/gb/food/pag/food35/food352.htm (accessed on October 17, 2005).

"FDA and Monosodium Glutamate (MSG)." U.S. Food and Drug Administration.
http://www.cfsan.fda.gov/~lrd/msg.html (accessed on October 17, 2005).

"Glutamate Facts, Information and On-Line Services." International Glutamate Information Service.
http://www.glutamate.org (accessed on October 17, 2005).

"L-glutamic Acid, Sodium Salt." J. T. Baker.
http://www.jtbaker.com/msds/englishhtml/g3975.htm (accessed on October 17, 2005).

N,N-Diethyl-3-Methyl-benzamide

OVERVIEW

N,N-diethyl-3-methylbenzamide (en-en-dye-ETH-el-three-METH-el-ben-ZA-mid) is the most commonly used insect repellant in the world. It is probably better known by its common name of DEET. DEET is applied to the skin and clothing to ward off biting insects such as mosquitoes, ticks, chiggers, and fleas. Although it does not kill insects, DEET repels them for several hours after being applied.

N,N-diethyl-3-methylbenzamide is a colorless amber liquid with a faint odor. The compound can exist in any one of three isomers in which the two groups attached to the benzene ring are next to each other (ortho), separated by one carbon (meta), or opposite each other in the ring (para). Although all three isomers are effective as insect repellants, the meta isomer is more effective than the ortho and para isomers and constitutes about 95 percent of the commercial product.

DEET was discovered by researchers at the U.S. Department of Agriculture and patented by the U.S. Army in 1946

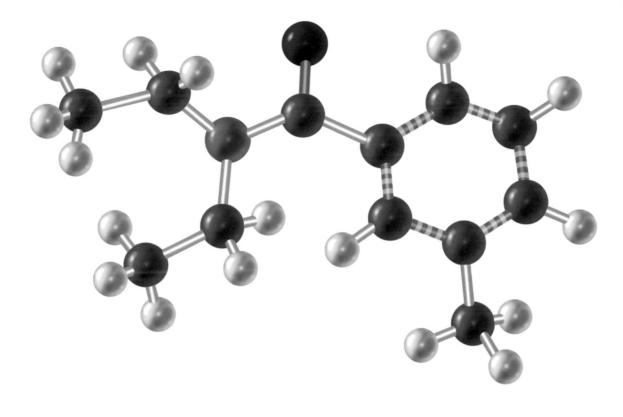

N,N-diethyl-3-methyl-benzamide. Red atom is oxygen; white atoms are hydrogen; black atoms are carbon; and blue atom is nitrogen. Gray stick indicates a double bond; striped sticks show a benzene ring.

PUBLISHERS RESOURCE GROUP

for use on military personnel working in insect-infested areas. It was not made available to the general public until 1957. Experts now regard DEET as the "gold standard" of insect repellants. It is available in a variety of formulations, including solutions, lotions, creams, gels, aerosol and pump sprays, and impregnated towelettes under names such as Hour Guard 8 and Hour Guard 12, DEET Plus, Sawyers Gold, Off (in many varieties), Ben's Backyard and Ben's Wilderness, Cutter (in many varieties), and Repel (in many varieties).

HOW IT IS MADE

DEET is made commercially by treating *m*-toluic acid ($C_6H_4CH_3COOH$) with thionyl chloride ($SOCl_2$). The product of this reaction, *m*-toluoyl chloride ($C_6H_4CH_3COOCl$) is then treated with diethylamine ($(C_2H_5)_2NH$) to obtain N,N-diethyl-3-methylbenzamide.

Interesting Facts

• The U.S. Environmental Protection Agency (EPA) has estimated that more than two hundred million people around the world use products containing N,N-diethyl-3-methylbenzamide.

COMMON USES AND POTENTIAL HAZARDS

The mechanism by which DEET repels insects is not completely understood. Current theories suggest that the compound blocks receptors on a mosquito's antennae that help it to locate carbon dioxide and lactic acid, given off in skin perspiration and the breath. Lacking these chemical clues, insects are unable to locate their prey (humans and other animals).

By warding off biting insects, DEET protects against the diseases they carry. Mosquitoes, for example, carry diseases such as malaria, one of the most serious diseases in the world, responsible for an estimated three millions deaths a year; encephalitis, an infection that causes inflammation and swelling of the brain; and West Nile virus, an organism that affects the central nervous system and poses a serious threat to both humans and other animals. Ticks carry Lyme disease, an infection spread by the deer tick that causes a skin rash, joint pain, and flu-like symptoms that can develop into a debilitating and permanent health problem if not treated early.

DEET is a very safe product when used as directed. It is absorbed rapidly through the skin, with up to 56 percent of the compound penetrating the skin in a six-hour period. Within twelve hours of application, DEET is metabolized by the liver and excreted in the urine. If inhaled or swallowed, however, DEET poses serious health hazards. It may cause headaches, seizures, and swelling of the brain, although these symptoms are very rare among users of the product. Products containing DEET are not recommended on babies under the age of two months, and are now required by the EPA

Words to Know

ISOMER One of two or more forms of a chemical compound with the same molecular formula, but different structural formulas and different chemical and physical properties.

to carry a warning to avoid spraying the product directly into the eyes or open wounds.

FOR FURTHER INFORMATION

"Basic Facts About DEET and DEET-Based Insect Repellents." Consumer Specialty Products Association. http://www.deet.com/deet_fact_sheet.htm (accessed on October 7, 2005).

"DEET." http://www.chem.ox.ac.uk/mom/InsectRepellents/DEET.htm (accessed on October 7, 2005).

Fradin, Mark S., M.D. "Mosquitoes and Mosquito Repellants." *Annals of Internal Medicine.* June 1, 1998: 931-940. Also available online at http://www.annals.org/cgi/content/full/128/11/931 (accessed on October 7, 2005).

Qiu H., H. W. Jun, and J. W. McCall. "Pharmacokinetics, formulation, and safety of insect repellent N,N-diethyl-3-methylbenzamide (DEET): a review." *Journal of the American Mosquito Control Association.* March 1998: 12-27. Also available online at http://www.ncbi.nlm.nih.gov/entrez/query.fcgi?cmd= Retrieve&db=PubMed&list_uids=98261685&dopt=Citation (accessed on October 7, 2005).

"Review of the Toxicology Literature for the Topical Insect Repellent Diethyl-m-toluamide (DEET)." Department of Health Toxicology Unit (United Kingdom). http://www.advisorybodies.doh.gov.uk/pdfs/reviewofdeet.pdf (accessed on October 7, 2005).

See Also Carbon Dioxide; Lactic Acid

Naphthalene

OVERVIEW

Naphthalene (NAF-thuh-leen) is a white crystalline volatile solid with a characteristic odor often associated with
mothballs. The compound sublimes (turns from a solid to a
gas) slowly at room temperature, producing a vapor that is
highly combustible. Naphthalene was first extracted from
coal tar in 1819 by English chemist and physician John Kidd
(1775-1851). Coal tar is a brown to black thick liquid formed
when soft coal is burned in an insufficient amount of air. It
consists of a complex mixture of hydrocarbons, similar to
that found in petroleum. Kidd's extraction of naphthalene
was of considerable historic significance because it demon-
strated that coal had other important applications than its use
as a fuel. It could also be utilized as the source of chemical
compounds with a host of important commercial and industrial uses. Naphthalene's chemical structure was determined
by the German chemist Richard August Carl Emil Erlenmeyer
(1825-1909). Erlenmeyer showed that the naphthalene molecule consists of two benzene molecules joined to each other.

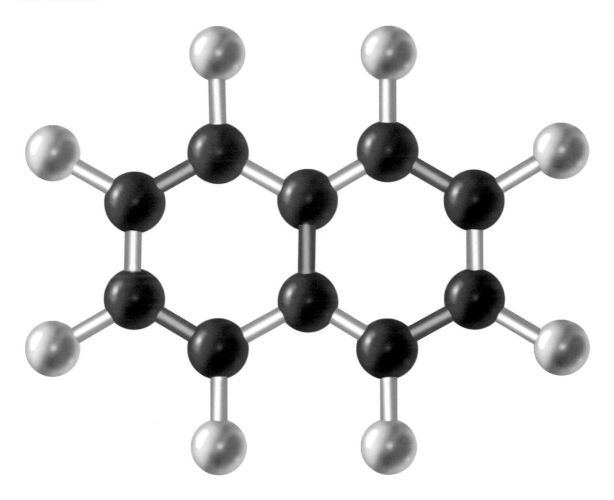

Naphthalene. White atoms are hydrogen and black atoms are carbon. Gray sticks indicate double bonds. PUBLISHERS RESOURCE GROUP

HOW IT IS MADE

Naphthalene occurs naturally in petroleum and coal tar. It is extracted from either by heating the raw material to a temperature of 200°C to 250°C (392°F to 482°F), producing a mixture of hydrocarbons known as middle oil. The middle oil is then distilled to separate its individual components from each other, one of which is naphthalene. The naphthalene produced in this process is purified by washing it in a strong acid and then in a sodium hydroxide solution and purified by steam distillation.

COMMON USES AND POTENTIAL HAZARDS

In 2000, about 100,000 metric tons (110,000 short tons) of naphthalene was produced in the United States. About 60

Interesting Facts

- The name "naphthalene" comes from the Persian word *naphtha*, meaning "oil" or "pitch."

- One use of naphthalene is as an insecticide. Nonetheless, one type of insect, the Formosan termite, uses naphthalene to build its nests. Scientists have not yet determined the source from which the termites get naphthalene or why they are immune to its otherwise toxic properties.

- Naphthalene becomes an environmental contaminant when it leaches into the soil. In 2003, however, scientists discovered bacteria capable of decomposing naphthalene in the soil, reducing the environmental hazards it poses.

- One of the most terrible weapons used during the Vietnam War of the 1960s and 1970s was a mixture of naphthalene and palmitate called napalm that ignited and burned at a very high temperature causing widespread destruction of plant life, structures, and human life.

percent of that amount was used for the synthesis of phthalic anhydride [$C_6H_4(CO)_2O$], a compound used as the starting point in the manufacture of a host of products, including a variety of dyes, resins, lubricants, insecticides, and a number of other products. At one time, naphthalene was used widely as a moth repellent in mothballs and moth flakes and in the manufacture of insecticides and fungicides. Demand for these applications has decreased, however, with the introduction of a class of compounds known as the chlorinated hydrocarbons. For example, one of those compounds, *p*-dichlorobenzene, is now the primary ingredient, rather than naphthalene, in mothballs and moth flakes.

Some other uses of naphthalene include:

- As an ingredient in the manufacture of the explosive known as smokeless gunpowder;

- In the manufacture of scintillation counters, devices used to detect and measure radioactivity;

Words to Know

DISTILLATION A process of separating two or more substances by boiling the mixture of which they are composed and condensing the vapors produced at different temperatures.

• In the production of lubricants;

• In the manufacture of various types of dyes and coloring agents;

• As an antiseptic and antihelminthic (a substance used to kill disease-causing worms);

• As a deodorant in toilets, diaper pails, and other settings; and

• As a preservative.

Naphthalene can be toxic by ingestion, inhalation, or absorption through the skin when humans are exposed to it in more than a passing way. In small amounts it may produce symptoms such as nausea, vomiting, headache, and fever. In larger amounts, it may induce more serious problems, including liver damage, blindness, coma, convulsions, and death. One medical problem that has been associated with exposure to large amounts of naphthalene is hemolytic anemia, a condition in which red blood cells are destroyed, resulting in fatigue, restlessness, jaundice, loss of appetite, and, in more serious cases, kidney failure.

FOR FURTHER INFORMATION

"Background and Environmental Exposures to Naphthalene, 1-Methylnaphthalene, and 2-Methylnaphthalene in the United States." Agency for Toxic Substances and Disease Registry. http://www.atsdr.cdc.gov/toxprofiles/tp67-c2.pdf (accessed on October 17, 2005).

Milius, Susan. "Termites Use Mothballs in Their Nests." *Science News* (May 2, 1998): 228.

"Naphthalene Chemical Backgrounder." National Safety Council. http://www.nsc.org/ehc/chemical/Naphthal.htm (accessed on October 17, 2005).

Naproxen

OVERVIEW

Naproxen (nah-PROK-sin) is a white to off-white odorless crystalline solid sold under a variety of commercial names, including Aleve®, Anaprox®, Bonyl®, Calosen®, Diocodal®, Naprosyn®, Naprelan®, Proxen®, and Veradol®. It is a nonsteroidal anti-inflammatory drug (NSAID) used as an analgesic (pain reliever) and antipyretic (fever-reducing compound). The compound is also available as a sodium salt, which is more readily absorbed in the gastrointestinal tract. Nonsteroidal anti-inflammatory drugs are compounds used to reduce pain, fever, and inflammation without the use of steroids (thus, nonsteroidal) by inhibiting the action of an enzyme needed to produce these results in the body. Other examples of NSAIDs are aspirin, ibuprofen, and prednisone.

HOW IT IS MADE

Naproxen is prepared by treating 2-methoxynaphthalene with a derivative of propanoic acid (CH_3CH_2COOH).

Naproxen. Red atoms are oxygen; white atoms are hydrogen; and black atoms are carbon. Striped sticks show benzene rings. PUBLISHERS RESOURCE GROUP

2-methoxynaphthalene has a structure very similar to that of naproxen and requires only the addition of the propanoyl group (-CH₂CH₂COOH) provided by propanoic acid. The reaction is fairly straightforward but includes one important consideration. Naproxen is a chiral compound,

CHEMICAL COMPOUNDS

Interesting Facts

• The U.S. Food and Drug Administration first approved the sale of naproxen as an over-the-counter medication (one that does not require a prescription) in 1994.

meaning that it can exist in one of two isomeric forms. The two forms are a "right-handed" form (designated as the R form) and a "left-handed" form (designated as the S form). As is the case with many organic compounds, the two chiral forms of naproxen have very different biological activities. Specifically, the S-form of naproxen is 28 times as effective as the R-form in achieving the analgesic, antipyretic, and anti-inflammatory results expected from the compound. Early methods for preparing naproxen resulted in the formation of a racemic mixture of R and S forms. A racemic mixture is one that contains both of the isomeric forms in which a compound can occur. The two parts of the racemic mixture, the R and S forms, then had to be separated from each other. Researchers have now developed a method of manufacture that produces essentially pure S naproxen, avoiding the time-consuming and costly process of separation previously required.

COMMON USES AND POTENTIAL HAZARDS

Naproxen is commonly used to treat the pain and stiffness caused by conditions such as arthritis, inflammation of the joints, gout, tendonitis, and bursitis. It achieves this result by inhibiting the activity of an enzyme called cyclooxygenase 2 (COX-2). COX-2 facilitates the production of a class of compounds in the body known as prostaglandins. Prostaglandins are produced when the body is injured, causing the pain and inflammation usually associated with an injury. Naproxen blocks the active site on a COX-2 molecule that catalyzes the formation of COX-2 molecules with the result: no COX-2; no prostaglandins; no pain and inflammation. Interestingly enough, a very similar enzyme called cyclooxygenase

Words to Know

CATALYST A material that increases the rate of a chemical reaction without undergoing any change in its own chemical structure.

ISOMER One of two or more forms of a chemical compound with the same molecular formula, but different structural formulas and different chemical and physical properties.

1 (COX-1) also exists in the body. COX-1, however, has a different set of functions that have nothing to do with the production of pain and inflammation. Since naproxen inhibits COX-2, but not COX-1, it is known as a selective inhibitor NSAID.

A number of relatively minor side effects have been reported for naproxen, including headache, dizziness, drowsiness, itching, ringing in the ears, and minor gastrointestinal discomfort that may include heartburn, abdominal pain, nausea, constipation, and diarrhea. More serious problems have also been reported, including gastrointestinal complications such as bleeding and ulceration (bleeding sores) of the stomach and intestines. In some cases, the drug has been responsible for the hospitalization of users.

Care needs to be taken when combining naproxen with other medications. Known adverse drug interactions can occur with aspirin, methotrexate, ACE inhibitors (for high blood pressure), furosemide, lithium, and warfarin (a blood thinner). An overdose of naproxen may cause dizziness, drowsiness, and gastrointestinal problems. High blood pressure, kidney failure, and coma may occur, but are rare.

FOR FURTHER INFORMATION

Buschmann, Helmut, et al. *Analgesics: From Chemistry and Pharmacology to Clinical Application.* New York: Wiley-VCH, 2002.

"Naproxen." World of Molecules. http://www.worldofmolecules.com/drugs/naproxen.htm (accessed on October 17, 2005).

Omudhome, Ogbru. "Naproxen Center." http://www.medicinenet.com/naproxen/article.htm (accessed on October 17, 2005).

"Product Monograph: Anaprox®; Anaprox® DS." Roche. http://www.rochecanada.com/pdf/Anaprox%20PM%20E.pdf (accessed on October 17, 2005).

OTHER NAMES:
Nicotinic acid;
3-pyridinecarboxylic
acid; vitamin B₃

FORMULA:
$C_6H_5NO_2$

ELEMENTS:
Carbon, hydrogen,
nitrogen, oxygen

COMPOUND TYPE:
Carboxylic acid
(organic)

STATE:
Solid

MOLECULAR WEIGHT:
123.11 g/mol

MELTING POINT:
236.6°C (457.9°F)

BOILING POINT:
Not applicable;
sublimes above its
melting point

SOLUBILITY:
Slightly soluble in
water, ethyl alcohol,
and ether

KEY FACTS

Niacin

OVERVIEW

Niacin (NYE-uh-sin) is a B vitamin (vitamin B₃) that is essential to cell metabolism. It occurs in two forms, nicotinic acid and nicotinamide, also called niacinamide. The only structural difference between the two compounds is that a hydroxyl group (-OH) in nicotinic acid is replaced by an amino group (-NH₂) group in nicotinamide. Lack of niacin causes a disease called pellagra. Pellagra was common throughout human history among poor people whose diet consisted almost entirely of corn products. Those corn products did not supply adequate amounts of niacin, causing symptoms such as diarrhea, scaly skin sores, inflamed mucous membranes, weakness, irritability, and mental delusions. In some cases, people with niacin deficiency develop reddish sores and rashes on their faces. Mental hospitals were full of people who seemed crazy, but who were actually suffering from a dietary deficiency. Thousands of people died from pellagra every year.

Nicotinic acid was first isolated by the Polish-American biochemist Casimir Funk (1884-1967) in 1912. At the time,

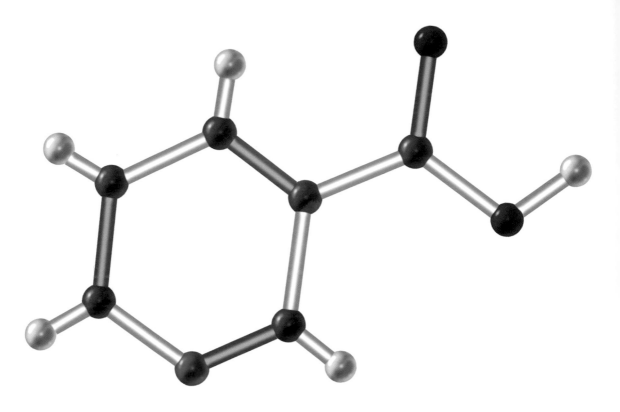

Niacin. Red atoms are oxygen; white atoms are hydrogen; black atoms are carbon; blue atom is introgen. Gray sticks indicate double bonds.

Funk was attempting to find a cure for another dietary disease known as beriberi. Since nicotinic acid had no effect on beriberi, he abandoned his work with that compound. It was left, then, to the Austrian-American physician Joseph Goldeberger (1874-1929) to find the connection between nicotinic acid and deficiency diseases. In 1915, Goldberger conducted a series of experiments with prisoners in a Mississippi jail and found that he could produce pellagra by altering their diets. He concluded that the disease was caused by the absence of some factor, which he called the P-P (for pellagra-preventative) factor. The chemical structure of that factor was then discovered in 1937 by the American biochemist Conrad Arnold Elvehjem (1901-1962), who cured the disease in dogs by treating them with nicotinic acid.

HOW IT IS MADE

Niacin is synthesized naturally in the human body beginning with the amino acid tryptophan. Tryptophan occurs naturally in a number of foods, including dairy products,

Interesting Facts

- Early investigators chose to use the name niacin rather than nicotinic acid because the later term sounds too much like "nicotine." They did not want to give the impression that a person can get the niacin they need by smoking tobacco, which contains nicotine.

- Some good sources of niacin include organ meats, such as kidney, liver, and heart; chicken; fish, such as salmon and tuna; milk and dairy products; brewer's yeast; dried beans and legumes; nuts and seeds; broccoli and asparagus; whole grains; dates; mushrooms; tomatoes; sweet potatoes; and avocados.

- Corn is also a good source of niacin. However, people with corn-based diets are at risk for developing pellagra. The reason for this contradiction is that the niacin in corn is chemically inactivated. Corn must be treated with an alkaline material to convert the niacin it contains to a free form that the body can use. Native Americans who treated their corn products with ley, limestone, wood ashes, or other alkaline materials did not suffer from pellagra.

beef, poultry, barley, brown rice, fish, soybeans, and peanuts. People whose diet consists mainly of corn products do not ingest adequate amounts of tryptophan, so their bodies are unable to make the niacin they need to avoid developing pellagra. It takes about 60 milligrams of tryptophan to produce 1 mg of niacin.

COMMON USES AND POTENTIAL HAZARDS

Niacin plays a number of essential roles in the body. It is necessary for cell respiration; metabolism of proteins, fats, and carbohydrates; the release of energy from foods; the secretion of digestive enzymes; the synthesis of sex hormones; and the proper functioning of the nervous system. It is also involved in the production of serotonin, an essential

Words to Know

METABOLISM Process that includes all of the chemical reactions that occur in cells by which fats, carbohydrates, and other compounds are broken down to produce energy and the compounds needed to build new cells and tissues.

MUCOUS MEMBRANES Tissues that line the moist inner lining of the digestive,

respiratory, urinary, and reproductive systems.

NEUROTRANSMITTER A chemical that carries nerve transmissions from one nerve cell to an adjacent nerve cell.

SUBLIME Change in state from a solid directly to a gas.

neurotransmitter in the brain. Niacin deficiency disorders occur as the result of an inadequate diet, consuming too much alcohol, and among people with certain types of cancer and kidney diseases. Physicians treat niacin deficiency diseases by prescribing supplements of 300 to 1,000 milligrams per day of the vitamin. Overdoses of niacin can cause a variety of symptoms, including itching, burning, flushing, and tingling of the skin.

FOR FURTHER INFORMATION

Eades, Mary Dan. *The Doctor's Complete Guide to Vitamins and Minerals.* New York: Dell, 2000.

"Niacin (Nicotinic Acid)." PDRHealth. http://www.pdrhealth.com/drug_info/nmdrugprofiles/ nutsupdrugs/nia_0184.shtml (accessed on October 20, 2005).

"Niacin Deficiency." *The Merck Manual.* http://www.merck.com/mrkshared/mmanual/section1/ chapter3/3l.jsp (accessed on October 20, 2005).

"Nicotinic Acid." Gondar Design Science. http://www.purchon.com/biology/nicotinic.htm (accessed on October 20, 2005).

OTHER NAMES:
r(S)-3-(1-methyl-2-pyr-
rolidinyl)pyridine;
1-methyl-2-(3-pyridyl)-
pyrrolidine

FORMULA:
$C_5H_4NC_4H_7NCH_3$

ELEMENTS:
Carbon, hydrogen,
nitrogen

COMPOUND TYPE:
Alkaloid (organic)

STATE:
Liquid

MOLECULAR WEIGHT:
162.23 g/mol

MELTING POINT:
$-79°C$ ($-110°F$)

BOILING POINT:
247°C (477°F)

SOLUBILITY:
Miscible with water;
very soluble in ethyl
alcohol, ether, and
chloroform

KEY FACTS

Nicotine

OVERVIEW

Nicotine (NIK-uh-teen) is a thick, colorless to yellow, oily liquid with a bitter taste that turns brown when exposed to air. It occurs in high concentrations in the leaves of tobacco plants and in lower concentrations in tomatoes, potatoes, eggplants, and green peppers. Nicotine gets its name from the tobacco plant, *Nicotiana tabacum*, which, in turn, was named in honor of the French diplomat and scholar Jean Nicot (1530-1600), who introduced the use of tobacco to Paris. Nicotine's correct chemical structure was determined in 1843 by the Belgian chemist and physicist Louise Melsens (1814-1886) and the compound was first synthesized by the research team of A. Pictet and A. Rotschy in 1904.

HOW IT IS MADE

Nicotine is extracted by soaking the stems and leaves of the tobacco plant in water for about twelve hours. After that period of time, the nicotine in the tobacco has dissolved in

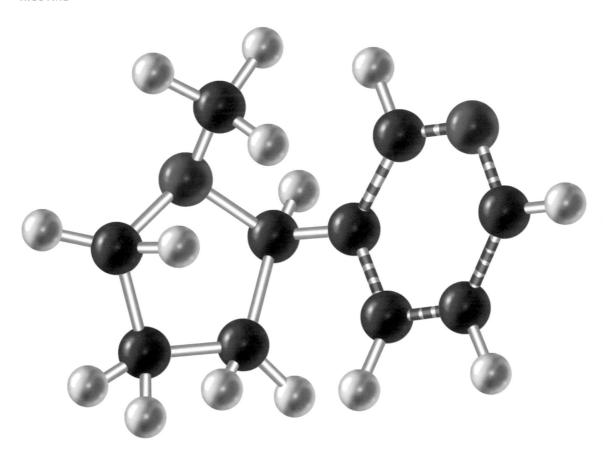

Nicotine. White atoms are hydrogen; black atoms are carbon; and blue atoms are nitrogen. PUBLISHERS RESOURCE GROUP

the water and can be extracted in a variety of ways. In one process, the water solution of nicotine is mixed with ether or chloroform, in which the nicotine is more soluble. The nicotine moves from the water layer to the ether or chloroform layer, from which it can be removed by evaporation.

COMMON USES AND POTENTIAL HAZARDS

The best-known application of nicotine is in tobacco products used for smoking and chewing. The actual nicotine content of tobacco products varies considerably, but, on average, ranges from about 15 to 25 milligrams per cigarette. Nicotine is also available in a number of products designed to help people stop smoking, such as nicotine gums and nicotine patches.

Nicotine was often used by farmers and gardeners as an insecticide and a fumigant in the past. Perhaps the best

Interesting Facts

- It takes only about seven seconds after nicotine is ingested before the chemical reaches the human brain.

- Nicotine is one of the most widely used addictive drugs in the United States.

known of these in the United States was an insecticide known as Black Leaf 40, a 40 percent solution of nicotine sulfate in water. The use of Black Leaf 40 and most other nicotine-containing insecticides has, to a large extent, been discontinued because of the toxic nature of the compound. The risk it posed to human users was greater than its value as an agricultural product.

Nicotine is a highly addictive substance. For that reason, people have difficulty stopping smoking or chewing tobacco products even when they recognize the health hazards posed by the compound. Smokers depend on nicotine to give them a burst of energy, since it stimulates the heart rate and quickens blood flow. Once a person becomes addicted to the use of nicotine, it requires larger doses of the compound to produce comparable effects. A 1998 U.S. government report issued by then-Surgeon General C. Everett Koop found that the addictive properties of nicotine are comparable to those of heroin and cocaine.

When a person uses tobacco, nicotine is quickly absorbed through respiratory tissues, the skin, and the gastrointestinal tract. The actual amount of nicotine absorbed by the body depends on a number of factors, including the type of tobacco being smoked and the presence or absence of a filter on the cigarette. After entering the body, nicotine flows through the bloodstream and across the blood brain barrier. Levels of the stimulating hormone adrenaline increase, as do blood sugar levels, respiration rates, blood pressure, and heart rate. Nicotine can make small arteries constrict, putting strain on the heart and raising blood pressure. If a person already has clogged arteries, this effect may cause heart pain (angina) or a heart attack.

Words to Know

MISCIBLE Able to be mixed; especially applies to the mixing of one liquid with another.

NEUROTRANSMITTER A chemical that carries nerve transmissions from one nerve cell to an adjacent nerve cell.

Although nicotine is a stimulant, it may induce muscle relaxation, depending on the user's physical state. It has also been shown to decrease one's appetite, speed up metabolism, and increase levels of dopamine, a mood-altering chemical in the brain that induces feelings of pleasure. Low levels of dopamine play a role in the development of Parkinson's disease. Research has shown that smokers, with higher levels of dopamine, have a reduced risk of the disease. Women who are pregnant are advised not to use any product containing nicotine. Nicotine in any form is harmful to an unborn child. It rapidly crosses the placenta and enters the fetus's body.

Nicotine is a highly toxic poison, which explains its former popularity as a pesticide. In high doses, it can be lethal. Low doses of nicotine can cause dizziness, nausea, and vomiting. Symptoms of acute nicotine poisoning may include a burning sensation in the mouth, more severe nausea and vomiting, diarrhea, heart palpitations, fluid in the lungs, seizures, coma, and death. People who smoke while receiving nicotine replacement therapy are at risk of nicotine poisoning.

FOR FURTHER INFORMATION

"Acute Nicotine Poisoning." *Mosby's Medical, Nursing, and Allied Health Dictionary.* 5th edition. St. Louis: Mosby, 1998.

"Facts about Nicotine and Tobacco Products." National Institute on Drug Abuse.
http://www.drugabuse.gov/NIDA_Notes/NNVol13N3/tearoff.html (accessed on October 20, 2005).

"Nicotine." International Labour Organization.
http://www.ilo.org/public/english/protection/safework/cis/products/icsc/dtasht/_icsc05/icsc0519.htm (accessed on October 20, 2005).

"Nicotine (Black Leaf 40) Chemical Profile 4/85." Pesticide Management Education Program, Cornell University. http://pmep.cce.cornell.edu/profiles/insect-mite/ mevinphos-propargite/nicotine/insect-prof-nicotine.html (accessed on October 20, 2005).

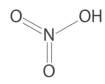

OTHER NAMES:
Aqua fortis;
engraver's acid;
azotic acid

FORMULA:
HNO$_3$

ELEMENTS:
Hydrogen, nitrogen,
oxygen

COMPOUND TYPE:
Inorganic acid

STATE:
Liquid

MOLECULAR WEIGHT:
63.01 g/mol

MELTING POINT:
−41.6°C (−42.9°F)

BOILING POINT:
83°C (180°F);
decomposes

SOLUBILITY:
Miscible with water;
decomposes in ethyl
alcohol; reacts
violently with most
organic solvents

Nitric Acid

OVERVIEW

Nitric acid (NYE-trik AS-id) is a colorless to yellowish liquid with a distinctive acrid (biting), suffocating, or choking odor. The acid tends to decompose when exposed to light, producing nitrogen dioxide (NO$_2$), itself a brownish gas. The yellowish tinge often observed in nitric acid is caused by the presence of small amounts of the nitrogen dioxide. Nitric acid is one of the strongest oxidizing agents known and attacks almost all metals with the notable exceptions of gold and platinum.

Nitric acid has been known to scholars for many centuries. Probably the earliest description of its synthesis occurs in the writings of the Arabic alchemist Abu Musa Jabir ibn Hayyan (c. 721–c. 815), better known by his Latinized name of Geber. The compound was widely used by the alchemists, although they knew nothing of its chemical composition. It was not until the middle of the seventeenth century that an improved method for making nitric acid was invented by German chemist Johann Rudolf Glauber (1604–1670). Glauber produced the acid by adding concentrated sulfuric acid (H$_2$SO$_4$)

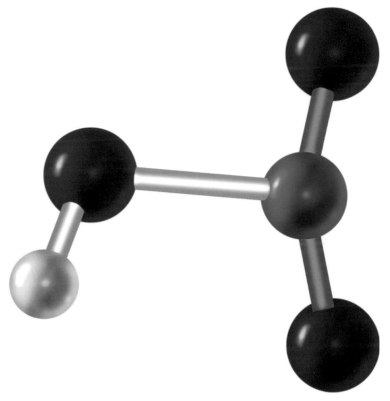

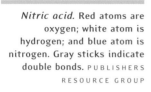

Nitric acid. Red atoms are oxygen; white atom is hydrogen; and blue atom is nitrogen. Gray sticks indicate double bonds. PUBLISHERS RESOURCE GROUP

to saltpeter (potassium nitrate; KNO_3). A similar method is still used for the laboratory preparation of nitric acid, although it has little or no commercial or industrial value.

The chemical nature and composition of nitric acid were first determined in 1784 by the English chemist and physicist Henry Cavendish (1731-1810). Cavendish applied an electric spark to moist air and found that a new compound—nitric acid—was formed. Cavendish was later able to determine the acid's chemical and physical properties and its chemical composition. The method of preparation most commonly used for nitric acid today was one invented in 1901 by the Russian-born German chemist Friedrich Wilhelm Ostwald (1853-1932). The Ostwald process involves the oxidation of ammonia over a catalyst of platinum or a platinum-rhodium mixture.

Today, nitric acid is one of the most important chemical compounds used in industry. It ranks number thirteen among all chemicals produced in the United States each year. In 2005, about 6.7 million metric tons (7.4 million short tons) of the compound were produced in the United States.

Interesting Facts

- Alchemists called nitric acid *aqua fortis*, a term that means "strong water."

- Nitric acid is a component of acid rain, a form of pollution that results when substances such as nitrogen oxides react with water, oxygen, and other chemicals in the atmosphere.

HOW IT IS MADE

Although several methods for the preparation of nitric acid are theoretically available, only one finds much commercial use: the direct oxidation of ammonia, an updated and improved version of the traditional Ostwald process. In this method, ammonia is heated and reacted with air over a catalyst, most commonly a mixture of rhodium and platinum metals. That reaction results in the formation of nitric oxide (NO), which is then converted to nitrogen dioxide (NO_2). The nitrogen dioxide reacts with water to form nitric acid.

COMMON USES AND POTENTIAL HAZARDS

The most common use for nitric acid is in the manufacture of ammonium nitrate, which, in turn, is used primarily as a fertilizer. About three-quarters of all nitric acid produced in the United States is used in fertilizers. The second most important application, accounting for about 10 percent of all nitric acid produced, is in the production of adipic acid [$COOH(CH_2)_3COOH$], used in the manufacture of nylon, polyurethanes, and other synthetic plastics. Nitric acid is also used to make a variety of metal nitrates and for the cleaning of metals. Small amounts of the compound are used for a variety of other applications, including:

- In the manufacture of explosives and fireworks;

- As a laboratory reagent in commercial, industrial, and academic research laboratories;

- In the processing of nuclear fuels;

- For the etching of metals; and

- In the manufacture of certain types of dyes.

Words to Know

ALCHEMY An ancient field of study from which the modern science of chemistry evolved.

CATALYST A material that increases the rate of a chemical reaction without undergoing any change in its own chemical structure.

MISCIBLE Able to be mixed; especially applies to the mixing of one liquid with another.

OXIDATION A chemical reaction in which oxygen reacts with some other substance or, alternatively, in which some substance loses electrons to another substance, the oxidizing agent.

Nitric acid is a highly toxic material. It attacks and destroys skin and other tissues, leaving a distinctive yellow scar caused by the destruction of proteins in the skin or tissue. If swallowed, inhaled, or spilled on the skin, it can cause a number of effects, including severe corrosive burns to the mouth, throat, and stomach; severe irritation or burning of the upper respiratory system, including nose, mouth, and throat; damage to the lungs; severe breathing problems; and burns to the eye surface, conjunctivitis, and blindness. In the most severe cases, the acid can cause death.

FOR FURTHER INFORMATION

"Concentrated Nitric Acid (70%)." International Chemical Safety Cards.
http://www.cdc.gov/niosh/ipcsneng/neng0183.html (accessed on October 20, 2005).

"Material Safety Data Sheet." Hill Brothers Chemical Company.
http://hillbrothers.com/msds/pdf/nitric-acid.pdf (accessed on October 20, 2005).

"Nitric Acid." Greener Industry.
http://www.uyseg.org/greener_industry/pages/nitric_acid/1nitricAcidAP.htm (accessed on October 20, 2005).

"Nitric Acid." Scorecard. The Pollution Information Site.
http://www.scorecard.org/chemical-profiles/summary.tcl?edf_substance_id=7697%2d37%2d2 (accessed on October 20, 2005).

See Also Ammonia; Ammonium Nitrate; Nitrogen Dioxide; Nitroglycerin

$$N\!\!=\!\!O$$

OTHER NAMES:
Nitrogen monoxide

FORMULA:
NO

ELEMENTS:
Nitrogen, oxygen

COMPOUND TYPE:
Nonmetallic oxide
(inorganic)

STATE:
Gas

MOLECULAR WEIGHT:
36.01 g/mol

MELTING POINT:
−163.6°C (−262.5°F)

BOILING POINT:
−151.74°C (−241.13°F)

SOLUBILITY:
Slightly soluble in
water

KEY FACTS

Nitric Oxide

OVERVIEW

Nitric oxide (NYE-trik OK-side) is a sweet-smelling, color-less gas that can be liquefied to make a bluish liquid and frozen to produce a bluish-white snow-like solid. It is one of five oxides of nitrogen, the others being nitrous oxide (N_2O), nitric oxide (NO), dinitrogen trioxide N_2O_3), and nitrogen dioxide (NO_2). Nitric oxide was first discovered in 1620 by Flemish physician and alchemist Jan Baptista van Helmont (1580-1635 or 1644).

Nitric oxide is used in the production of nitric acid, ammonia, and other nitrogen-containing compounds. It is also formed as a byproduct of the combustion of coal and petroleum products. As such, it is a major contributor to air pollution.

HOW IT IS MADE

Nitrogen and oxygen are the two most abundant gases in the atmosphere. Since both elements are relatively inactive,

Nitrogen monoxide. Red atom is oxygen and blue atom is nitrogen. PUBLISHERS RESOURCE GROUP

they do not combine with each other under normal circumstances. However, the energy provided by lightning strikes causes the reaction of the two elements, producing nitric oxide.

$$N_2 + O_2 \rightarrow 2NO$$

Another source of the energy needed for this reaction is the combustion of coal and oil products used for human activities. For example, the combustion of gasoline in an internal combustion engine produces temperatures in excess of 2,000°C (3600°F). These temperatures are sufficient to bring about the reaction between nitrogen and oxygen in a vehicle engine, resulting in the formation of nitric oxide as one product of gasoline combustion. That nitric oxide then passes into the atmosphere and becomes a major component of air pollution.

COMMON USES AND POTENTIAL HAZARDS

The most important industrial use of nitric oxide is in the preparation of other nitrogen-containing compounds, especially nitrogen dioxide (NO_2), nitric acid (HNO_3), and nitrosyl chloride ($NOCl$). It also finds some application in the bleaching of rayon (a synthetic, or artificially created, fabric) and as a polymerization inhibitor with certain compounds such as propylene and methyl ether. Such compounds have a tendency to react with each other to form large, complex molecules known as polymers.

Nitric oxide is considered an environmental pollutant. It oxidizes readily to form nitrogen dioxide, which, in turn,

Interesting Facts

- In 1992, *Science* magazine named nitric oxide "Molecule of the Year" after scientists discovered that it had several important functions in the body.

- Ferid Murad (1936-), Robert Furchgott (1916-), and Louis Ignarro (1941-) shared the 1998 Nobel Prize in Physiology or Medicine for their discovery of the role played by nitric oxide in the body's nervous system.

reacts with moisture in the air to form nitric acid, a component of acid rain. Acid rain is thought to be responsible for a number of environmental problems, including damage to buildings, destruction of trees, and the death of aquatic life. The nitrogen dioxide produced from nitric oxide is also a primary component of photochemical smog, a hazardous haze created by a mixture of pollutants in the presence of sunlight.

Even though it is toxic in the environment, nitric oxide plays several important roles in the human body. Nitric oxide is involved in the process by which messages are transmitted from one nerve cell to the next. It also regulates blood flow by triggering the smooth muscles surrounding blood vessels to relax. This action increases blood flow and lowers blood pressure. Nitric oxide also prevents the formation of blood clots, which can break off and travel to the heart or brain, increasing the risk of heart attack or stroke.

During sexual arousal, nitric oxide increases blood flow to the penis, leading to an erection in a man. The drug Viagra stimulates erections by enhancing the flow of nitric oxide in the penis.

Finally, nitric oxide plays a role in memory and learning. A deficiency of the compound appears to be related to the development of learning problems. On the other hand, an excess of nitric oxide has been implicated in the development of certain diseases, such as Huntington's chorea, an inherited disorder characterized by unusual body movements

Words to Know

POLYMERIZATION The process of creating a polymer, a compound consisting of very large molecules made of one or two small repeated units called monomers.

and memory loss, and Alzheimer's disease, a progressive disorder that results in memory loss.

When used to treat a medical condition, nitric oxide is usually administered in the form of a solid or liquid medicine that decomposes in the body, releasing the compound. For example, the drug nitroglycerin is used to treat heart problems. When it enters the bloodstream, nitroglycerin begins to break down, releasing nitric oxide. The nitric oxide causes smooth muscle cells in the heart to relax, relieving the symptoms of angina, chest pain caused by an inadequate flow of blood to the heart. Other types of drugs produce nitric oxide to inhibit the buildup of fatty deposits in blood vessels, which can lead to heart attack and stroke. Patients with pulmonary hypertension, a condition in which the vessels that supply blood to the lungs are constricted, preventing normal oxygen flow, are sometimes given an inhaler with a mixture of nitric oxide and air to open blood vessels to the lungs.

In spite of its many benefits, nitric oxide may also be a health hazard. If inhaled in excessive amounts, it may replace oxygen in the lungs, leading to asphyxia, suffocation resulting from an insufficient supply of oxygen. Research suggests that exposure to low concentrations of the gas over long periods of time may result in lung disease, emphysema, and chronic bronchitis.

FOR FURTHER INFORMATION

Butler, A. R., and R. Nicholson. *Life, Death and Nitric Oxide.* London: Royal Society of Chemistry, 2003.

"Gas Data." Air Liquide.
http://www.airliquide.com/en/business/products/gases/gasdata/index.asp?GasID=44 (accessed on October 20, 2005).

"Nitric Oxide." Reproductive and Cardiovascular Disease Research Group. http://www.sgul.ac.uk/depts/immunology/~dash/no/ (accessed on October 20, 2005).

"Nitrogen Oxides." International Programme on Chemical Safety. http://www.inchem.org/documents/ehc/ehc/ehc188.htm# SubSectionNumber:2.1.1 (accessed on October 20, 2005).

Stanley, Peter. "Nitric Oxide." Biological Sciences Review (April 2002): 18-20.

See Also Nitric Acid; Nitrogen Dioxide

OTHER NAMES:
Dinitrogen tetroxide;
nitrogen peroxide

FORMULA:
NO_2

ELEMENTS:
Nitrogen, oxygen

COMPOUND TYPE:
Nonmetallic oxide
(inorganic)

STATE:
Gas or liquid

MOLECULAR WEIGHT:
46.01 g/mol

MELTING POINT:
−11.2°C (11.8°F)

BOILING POINT:
21.15°C (70.07°F)

SOLUBILITY:
Reacts with water to
form nitric acid
(HNO_3) and nitrous
acid (HNO_2)

KEY FACTS

Nitrogen Dioxide

OVERVIEW

Nitrogen dioxide (NYE-truh-jin dye-OK-side) is a toxic reddish-brown gas or yellowish-brown liquid with a pungent, irritating odor. Above 21.15°C (70.07°F), it exists as the reddish-brown gas. Below that temperature, it becomes the yellowish-brown liquid. When liquified under pressure, it forms a fuming brown liquid. The brown liquid is actually a mixture of nitrogen dioxide and dinitrogen tetroxide (N_2O_4), a dimeric form of nitrogen dioxide. A dimer is a molecule that consists of two identical molecules combined with each other. When cooled below −11.2°C (11.8°F), the liquid freezes to form a colorless crystalline solid that consists almost entirely of the dimeric form, N_2O_4.

Small amounts of nitrogen dioxide are present naturally in the atmosphere as the result of lightning strikes, volcanic action, forest fires, and bacterial action on dead plants and animals. Much larger amounts are present because of human activities, primarily the combustion of fossil fuels, such as coal and petroleum products.

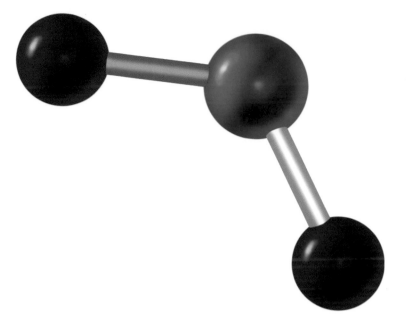

Nitrogen dioxide. Red atoms are oxygen and blue atom is nitrogen. Gray stick indicates a double bond. PUBLISHERS RESOURCE GROUP

HOW IT IS MADE

Nitrogen dioxide occurs naturally when nitric oxide (NO) is oxidized in the atmosphere. Nitric oxide forms naturally whenever sufficient energy is available to make possible the reaction between nitrogen and oxygen, the two primary components of the atmosphere. The formula for this reaction is

$$N_2 + O_2 \rightarrow 2NO$$

Nitric oxide, in turn, reacts readily with oxygen to form nitrogen dioxide. This reaction can be shown as

$$2NO + O_2 \rightarrow 2NO_2.$$

The amount of nitrogen dioxide produced naturally is so small that the gas's distinctive brown color is unnoticeable.

Such is not the case, however, in situations where nitric oxide and nitrogen dioxide are produced by human activities. The burning of coal or oil in plants that generate electricity and the combustion of gasoline in automobiles and trucks produce significant quantities of nitric oxide, which rapidly oxidizes to form nitrogen dioxide in the atmosphere. In such cases, sufficient nitrogen dioxide may be present to produce the yellowish hazy condition associated with smog.

Commercially, nitrogen dioxide is made by one of two processes, the decomposition of nitric acid (HNO_3) or the

Interesting Facts

- Scientists often use the formula NOx to refer to the oxides of nitrogen that occur in polluted air. The two most important of those oxides are nitric oxide (NO) and nitrogen dioxide.

oxidation of ammonia (NH_3) gas. The gas can also be produced on a smaller scale or in the laboratory by a number of methods that involve the decomposition of the nitrate ion (NO_3). For example, heating lead(II) nitrate [$Pb(NO_3)_2$] results in the formation of lead(II) oxide (PbO), nitrogen dioxide, and oxygen.

COMMON USES AND POTENTIAL HAZARDS

The most important uses of nitrogen dioxide are in the manufacture of nitric and sulfuric acids, two of the most widely used inorganic acids. The compound is also used widely as an oxidizing agent and nitrating agent. An oxidizing agent is a substance that makes oxygen available to other compounds in order to bring about a particular reaction. A nitrating agent is one that provides the nitro group (-NO_2) in producing a new compound. As an example, adding nitrate groups to the organic compound known as toluene ($C_6H_5CH_3$) converts it into trinitrotoluene (TNT), a powerful explosive. Nitrogen dioxide is also used in rockets, where it supplies the oxygen needed to burn the rocket fuel; as a catalyst in a number of industrial operations; in the bleaching of flour; and as a polymerization inhibitor in the manufacture of some plastics. A polymerization inhibitor is a substance that stops the formation of a polymer at some desired point in its production or use.

Nitrogen dioxide poses both safety and health hazards. As a strong oxidizing agent, it reacts readily with combustible materials, such as paper, cloth, and other organic matter to produce fires or explosions. It is also a toxic material, producing some biological effects at relatively low concentrations in the air. These effects include irritation of the eyes, nose, and throat; coughing; congestion; chest pain; and breathing

Words to Know

DIMER A molecule that consists of two identical molecules combined with each other.

POLYMER A compound consisting of very large molecules made of one or two small repeated units called monomers.

POLYMERIZATION INHIBITOR A substance that stops the formation of a polymer at some desired point in its production or use.

difficulties. The gas is sometimes referred to as an insidious agent because its effects may go unnoticed for several hours or days, during which time more serious damage may have occurred. This damage may include pulmonary edema, a condition in which the lungs begin to fill with fluid; cyanosis, a condition in which the lips and mucous membranes turn blue because of lack of oxygen; and a variety of heart problems. Long-term exposure to nitrogen dioxide may result in chronic health problems, such as hemorrhaging (blood loss), lung damage, emphysema, chronic bronchitis, and eventually death.

FOR FURTHER INFORMATION

"Cheminfo: Chemical Profiles Created by CCOHS." http://www.intox.org/databank/documents/chemical/nitrodix/cie748.htm (accessed on October 20, 2005).

Holgate, S. T., et al. *Air Pollution and Health.* New York: Academic Press, 1999.

"NIOSH Pocket Guide to Chemical Hazards: Nitrogen Dioxide." Occupational Health & Safety Administration. http://www.cdc.gov/niosh/npg/npgd0454.html (accessed on October 20, 2005).

"Nitrogen Oxides." International Programme on Chemical Safety. http://www.inchem.org/documents/ehc/ehc/ehc188.htm#SubSectionNumber:2.1.1 (accessed on October 20, 2005).

Patnaik, Pradyot. *Handbook of Inorganic Chemicals.* New York: McGraw-Hill, 2003, 648-651.

See Also Nitric Acid

KEY FACTS

Nitroglycerin

OVERVIEW

Nitroglycerin (nye-tro-GLIH-cer-in) is a pale yellow oily flammable liquid that is highly explosive. It is used primarily as an explosive by itself and as an ingredient in dynamite. Nitroglycerin also finds application in medicine as a vasodilator, a substance that causes blood vessels to relax and open up, allowing blood to flow more freely through them.

Nitroglycerin was first developed in 1847 by the Italian chemist Ascanio Sobrero (1812-1888). Sobrero used a method of synthesis that is still the primary means of producing nitroglycerin today. He added nitric acid to glycerol, then and now a popular skin lotion, with a small amount of concentrated sulfuric acid as catalyst. He initially called his discovery *pyroglycerin*. When he placed a trace amount of nitroglycerin on his tongue, Sobrero discovered that it "gives rise to a most pulsating, violent headache, accompanied by a great weakness of the limbs." He also discovered that the chemical was highly combustible and explosive. He considered the substance so dangerous that he warned against its

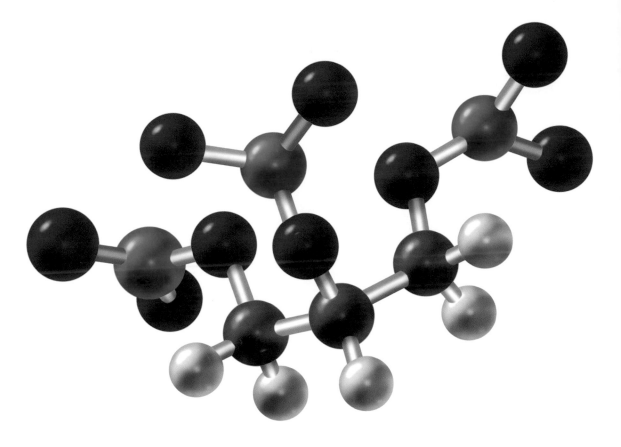

Nitroglycerin. Red atoms are oxygen; white atoms are hydrogen; black atoms are carbon; and blue atoms are nitrogen. Gray sticks indicate double bonds. PUBLISHERS RESOURCE GROUP

use and did not disclose his discovery to the world for more than a year.

In the 1860s, the Swedish chemist Alfred Nobel (1833-1896) developed a process for manufacturing nitroglycerin on a large scale. His company sold a combination of nitroglycerin and gunpowder called Swedish blasting oil, but the product proved far too dangerous to use. Several accidents involving the substance occurred. One explosion in Nobel's factory killed several people, including his brother Emil. To make a safer explosive, Nobel found a way to combine nitroglycerin with clay, a chemically inactive material. He called the combination dynamite. Dynamite soon became the explosive of choice in construction, demolition, and mining projects around the world.

Nitroglycerin's medical uses were first explored in detail by Sir Thomas Lauder Brunton (1844-1916) of the Royal Infirmary in Edinburgh, Scotland. In 1938, the U.S. Food

Interesting Facts

- Before Nobel began studying the explosive properties of nitroglycerin, people reportedly used the compound for everyday chores, such as polishing boots and greasing wagon wheels, and even as lamp oil. The consequences were often fatal.

- Although he spent most of his life developing and manufacturing explosives, Nobel was a humanitarian who wanted to see technology used for the benefit of society and the advancement of world peace. His handwritten will, although fiercely contested, provided for the creation of the Nobel Foundation. The foundation's primary responsibility is to award prizes in chemistry, physics, physiology or medicine, and peace every year. The first Nobel prizes were awarded in 1901. Today, a Nobel prize is regarded as the highest honor given in science. Each prize is worth over a million dollars in cash.

and Drug Administration first approved the use of nitroglycerin as a vasodilator for the treatment of heart problems.

HOW IT IS MADE

The preparation of nitroglycerin is a straightforward, but very delicate, chemical procedure. When nitric acid is added to glycerol, three nitrate groups ($-NO_2$) replace each of the three hydroxyl ($-OH$) groups on glycerol. This reaction occurs only if the acid and glycerol are moderately warm, at least at room temperature. But combining nitric acid, glycerol, and sulfuric acid (the catalyst) for the reaction generates heat. Too much heat causes the nitroglycerin being produced to explode. The resolution for this dilemma is to begin the reaction at room temperature, but then encase the reaction vessel in ice as soon as the reaction begins. The ice absorbs heat generated during the reaction, allowing the formation of nitroglycerin without it becoming too warm.

Words to Know

CATALYST A material that increases the rate of a chemical reaction without undergoing any change in its own chemical structure.

SYNTHESIS A chemical reaction in which some desired chemical product is made from simple beginning chemicals, or reactants.

COMMON USES AND POTENTIAL HAZARDS

Nitroglycerin's primary use is as an explosive, either by itself or as a component of dynamite. Today, it is marketed under more than 60 trade names, including Coro-Nitro®, Deponit®, GTN®, Nitroglin®, Nitrong®, Perglottal®, Reminitrol®, Sustac®, Tridil®, and Vasoglyn®. Nitroglycerin is classified as a high explosive, which means that it explodes very rapidly with a very large force. It is detonated ("set off") either by heat or by shock. Because of its very unstable character, it is usually transported at low temperatures (5°C to 10°C; 40°F to 50°F), at which it is more stable.

When used as a heart medication, nitroglycerin is administered in the form of a pill, patch, or intravenous solution. Nitroglycerin works in the body by releasing nitric oxide, a vasodilator. Vasodilators cause the smooth muscle surrounding blood vessels to relax, allowing the vessels to expand and improve the flow of blood. The drug is often taken when the pain of angina or a heart attack is first noticed.

People who come into contact with nitroglycerin in the workplace are at risk for a number of health problems. The compound is a skin, eye, and respiratory system irritant. It may cause nausea, vomiting, abdominal cramps, headache, mental confusion, delirium, sweating, a burning sensation on the tongue, paralysis, convulsions, and death.

FOR FURTHER INFORMATION

"Nitroglycerin." Imperial College of London, Department of Chemistry. http://www.ch.ic.ac.uk/rzepa/mim/environmental/html/nitroglyc_text.htm (accessed on October 20, 2005).

"Nitroglycerin: Dynamite for the Heart." *Chemistry Review* (November 1999): 28.

Rawls, Rebecca. "Nitroglycerin Explained." *Chemical & Engineering News* (June 10, 2002): 12.

"Why Is Nitroglycerin Explosive?" General Chemistry Online http://antoine.frostburg.edu/chem/senese/101/consumer/faq/nitroglycerin.shtml (accessed on October 20, 2005).

See Also Nitric Acid; Sulfuric Acid

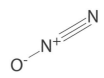

OTHER NAMES:
See Overview.

FORMULA:
N₂O

ELEMENTS:
Nitrogen; oxygen

COMPOUND TYPE:
Nonmetallic oxide
(inorganic)

STATE:
Gas

MOLECULAR WEIGHT:
44.01 g/mol

MELTING POINT:
−90.8°C (−131°F)

BOILING POINT:
−88.48°C (−127.3°F)

SOLUBILITY:
Slightly soluble in
water; soluble in
ethyl alcohol and
ether

KEY FACTS

Nitrous Oxide

OVERVIEW

Nitrous oxide (NYE-truss OX-side) is also known as dinitrogen oxide, dinitrogen monoxide, nitrogen monoxide, and laughing gas. It is a colorless, nonflammable gas with a sweet odor. Its common name of laughing gas is derived from the fact that it produces a sense of light-headedness when inhaled. The gas is widely used as an anesthetic, a substance that reduces sensitivity to pain and discomfort.

Nitrous oxide was probably first produced by the English chemist and physicist Robert Boyle (1627-1691), although he did not recognize the new compound he had found. Credit for the discovery of nitrous oxide is, therefore, usually given to the English chemist Joseph Priestley (1733-1804), who produced the gas in 1772 and named it "nitrous air." Other early names used for the gas include "gaseous of azote" (nitrogen) and "oxide of speton." The most complete experiments on the gas were conducted by the English chemist and physicist Sir Humphry Davy (1778-1829), who tested nitrous oxide on himself and his friends. He found that the gas could lessen

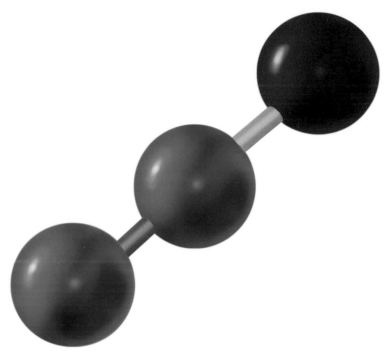

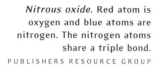
Nitrous oxide. Red atom is oxygen and blue atoms are nitrogen. The nitrogen atoms share a triple bond.

pain and discomfort and provided a sense of relaxation and well-being. Before long, doctors were making use of Davy's discovery by using nitrous oxide as an anesthetic.

The public found other uses for the gas as well. During the Victorian period in England, members of the upper class often held laughing gas parties at which people gathered to inhale nitrous oxide as a recreational drug, rather than for any therapeutic purpose. In the United States, the showman P. T. Barnum (1810-1891) created a sideshow exhibit in which people were invited to test the effects of inhaling nitrous oxide. After seeing a demonstration of this kind, the American dentist Horace Wells (1815-1848) first used nitrous oxide as an anesthetic on his patients.

In 1868, the American surgeon Edmund Andrews (1824-1904) extended the use of nitrous oxide as an anesthetic for his surgical patients. He mixed the gas with oxygen to ensure that patients received enough oxygen while receiving the anesthetic. The gas is still widely used by dentists as a safe and relatively pleasant way of helping patients endure the discomfort of drilling and other dental procedures.

Interesting Facts

- Humphry Davy proposed the name laughing gas for nitrous oxide.

- In the United Kingdom, nitrous oxide is often used as an anesthetic for women about to give birth.

- In the 1830s, Samuel Colt (1814-1862), inventor of the Colt 45 revolver, toured North America, giving laughing gas demonstrations.

HOW IT IS MADE

The most common commercial method of producing nitrous oxide involves the controlled heating of ammonium nitrate (NH_4NO_3). The compound decomposes to form nitrous oxide and water. The reaction is essentially the same one originally used by Priestley in 1772. Although an efficient means of producing the gas, the reaction must be carried out with extreme care as ammonium nitrate has a tendency to decompose explosively when heated. Nitrous oxide can also be produced by the decomposition of nitrates (compounds containing the NO_3 radical), nitrites (compounds containing the NO_2) radical, or nitriles (compounds containing the CH^- radical).

COMMON USES AND POTENTIAL HAZARDS

Nitrous oxide is best known and most widely used as an anesthetic. Its use is limited primarily to dental procedures and minor surgeries. Dentists favor nitrous oxide as an anesthetic because the gas does not make patients completely unconscious and does not require an anesthesiologist to administer it. Nitrous oxide works as an anesthetic by blocking neurotransmitter receptors in the brain, preventing pain messages from being transmitted.

Nitrous oxide is also used as a fuel additive in racing cars, in which case it is often referred to as nitro. The gas is injected into the intake manifold where it mixes with air and fuel vapors. Since it breaks down at the high temperatures in the car's engine, it provides additional oxygen to increase the

Words to Know

ANESTHETIC A substance that reduces sensitivity to pain and discomfort.

NITRATE A compound that includes the radical consisting of one nitrogen atom and three oxygen atoms (NO_3).

NITRITE A compound that includes the radical consisting of one nitrogen atom and two oxygen atoms (NO_2).

RADICAL A group of atoms bonded together that act like a single entity in chemical reactions.

efficiency with which the fuel burns. During World War II, pilots used nitrous oxide for a similar purpose in their airplanes.

Some additional uses of nitrous oxide include:

- As a propellant in food aerosols;
- For the detection of leaks;
- As a packaging gas for potato chips and other snack foods, preventing moisture from making the product become stale;
- In the preparation of other nitrogen compounds; and
- As an oxidizing agent for various industrial processes.

Nitrous oxide is safe to use in moderate amounts under controlled conditions. Some people use the compound as a recreational drug, however, hoping to get a "high" from inhaling it. One risk of this practice is that the inhalation of nitrous oxide may reduce the amount of oxygen a person receives. Also, some long-term health effects, such as anemia (low red blood cell count) and neuropathy (damage to the nerves), have been associated with excessive use of the compound. The use of nitrous oxide for recreational purposes is a crime in some states.

FOR FURTHER INFORMATION

"Gas Data: Nitrous Oxide." Air Liquide. http://www.airliquide.com/en/business/products/gases/gasdata/index.asp?GasID=55 (accessed on October 20, 2005).

Neff, Natalie. "No Laughing Matter." *Auto Week* (May 19, 2003): 30.

"Nitrogen Oxide." Center for Advanced Microstructures and Devices, Louisiana State University. http://www.camd.lsu.edu/msds/n/nitrous_oxide.htm#Synonyms (accessed on October 20, 2005).

"Occupational Safety and Health Guideline for Nitrous Oxide." Occupational Safety and Health Administration. http://www.osha.gov/SLTC/healthguidelines/nitrousoxide/recognition.html (accessed on October 20, 2005).

Pae, Peter. "Sobering Side of Laughing Gas." *Washington Post* (September 16, 1994): B1.

See Also Ammonium Nitrate

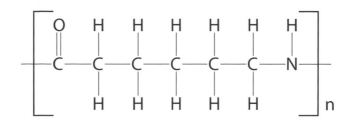

FORMULA:
Nylon 6:
-[-CO(CH$_2$)$_5$NH-]-$_n$;
Nylon 66:
-[-CO(CH$_2$)$_4$
CO-NH(CH$_2$)$_6$NH-]-$_n$

ELEMENTS:
Carbon, hydrogen,
oxygen, nitrogen

COMPOUND TYPE:
Polymer (organic)

STATE:
Solid

MOLECULAR WEIGHT:
Very large

MELTING POINT:
Nylon 6: 223°C
(433°F); Nylon 66:
265°C (509°C)

BOILING POINT:
Decomposes above
melting point

SOLUBILITY:
Insoluble in water and
most organic
solvents; soluble in
strong acids

Nylon 6 and Nylon 66

OVERVIEW

The term nylon is used to describe a family of organic polymers called the polyamides that contain the amide (-CONH) group. The members of the family are distinguished from each other by a numbering system indicating the chemical composition of the polymer molecule. The two most important nylons are nylon 6 and nylon 66 which, between them, account for nearly all of the nylon produced in the United States. Other nylons that are produced in much smaller amounts include nylon 11, nylon 12, nylon 46, and nylon 612.

The polyamides are thermoplastic polymers. The term "thermoplastic" means that the polymer can be repeatedly melted and hardened by alternate heating and cooling. By contrast, certain other types of polymers, known as thermosetting plastics, can not be re-melted once they have hardened.

Nylon was invented in 1935 by Wallace Carothers (1896-1937), an employee of the DuPont Chemical Corporation at the time. Carothers was searching for a synthetic substitute

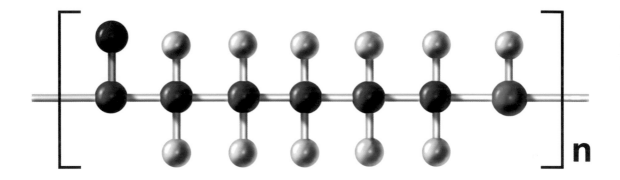

Nylon 6. Red atom is oxygen; white atoms are hydrogen; black atoms are carbon; blue atoms are nitrogen; gray stick shows a single bond.
PUBLISHERS RESOURCE GROUP

of silk because natural supplies were insufficient to meet growing demands for that fiber. Using coal, water, and air as raw materials, Carothers developed a synthetic product that could be stretched into a fiber and that got stronger and silkier as it was stretched. Nylon first appeared in commercial products in 1938, first in toothbrushes with nylon bristles, and later in women's stockings. During World War II, the U.S. military found a number of uses for the compound, especially the manufacture of parachutes. Today, nylon is used in a very wide array of products.

HOW IT IS MADE

Nylon 66 is still made commercially by the procedure originally developed by Carothers in 1935. The process is initiated with the reaction between adipic acid ($HOOC(CH_2)_4 COOH$) and hexamethylenediamine ($H2N(CH2)6NH2$), which results in the formation of the monomer ($-[-CO(CH_2)_4- CONHCH_2)_6-]-$) from which the polymer develops. That monomer has active groups at both ends of the molecule, allowing it to react with others of its kind to form long chains that make up the polymer. Methods for preparing the other polyamides can differ quite significantly from the Carothers procedure. For example, nylon 6 is made by the polymerization of aminocaproic acid ($H_2N(CH_2)_5COOH$). The polymer formed in this reaction also has blocks of methylene ($-CH_2$) groups joined by amide (CONH) groups, although the specific structure is somewhat different than it is for nylon 66.

COMMON USES AND POTENTIAL HAZARDS

Production of nylon 6 and nylon 66 for 2006 in the United States was estimated to be about 1.6 million kilograms (3.5 mil-

Interesting Facts

- The term nylon has never been trademarked. The name itself has no specific meaning. The -on suffix was chosen, however, to suggest a comparison with other fibers, such as cotton and rayon.

- DuPont originally thought of calling nylon No-Run. They abandoned that plan, however, when they realized that nylon stockings do, in fact, run.

lion pounds). About three-quarters of that production is used for the manufacture of fibers used in industrial operations and for textiles. Fabrics made from nylon are strong, resistant to abrasion, resistant to alkaline materials, and capable of taking most dyes. The clothing industry uses nylon fibers to make a variety of fabric types, including fleece, velvet, satin, taffeta, lace, and seersucker. Nylon clothing is lightweight, silky, attractive and smooth. Since it does not absorb moisture or odors readily, it is in demand for the production of athletic clothing. Nylon is also used in the manufacture of home textiles, such as furniture covering. Nylon rope is twice as strong as rope made from natural fibers such as hemp and weighs less.

About 15 to 20 percent of all nylon produced in the United States goes to the manufacture of molded plastics, such as those used in automobile and truck parts, housing for electrical devices, and consumer articles. The next largest application for nylon is in tubing, pipes, films, and coatings. Some specific articles made from nylon 6 and nylon 66 include the following:

- Reinforcing cord for tires;
- Fuel tanks for automobiles;
- Tennis rackets;
- Fishing nets and lines;
- Towlines for gliders;
- Women's stockings;
- Sutures for surgical procedures;

Words to Know

POLYAMIDE An organic polymer that contains the amide (-CONH) group.

POLYMER A compound consisting of very large molecules made of one or two small repeated units called monomers.

THERMOPLASTIC Capable of being heated so that it can be reshaped, then cooled so that it hardens.

THERMOSETTING Capable of being heated so that it hardens.

- Bristles for toothbrushes, hairbrushes, and paint brushes;
- Parachutes;
- Pen tips; and
- Artificial turf for athletic fields.

One of the most important uses for nylon historically was the manufacture of women's stockings. Silk stockings were an essential item of a well-dressed woman's ensemble in the 1920s and 1930s. They made a woman's legs look smooth and sleek, and they were the ideal accompaniment to short skirts and high heels. In the 1920s, women spent millions of dollars on silk stockings. The DuPont company began the search for a silk substitute because it hoped to develop a product that could be used to make stockings that looked and felt as if they were made of silk, but that cost less. Nylon stockings became commercially available in 1939, and by 1941, some sixty million pairs had been sold. Both nylon and silk stockings were difficult to get during World War II because the fibers were being used for military applications. As soon as the war was over, however, women began buying nylon stockings in large quantities once again.

FOR FURTHER INFORMATION

Hermes, Matthew E. *Enough for One Lifetime: Wallace Carothers, Inventor of Nylon.* Philadelphia: Chemical Heritage Foundation, 1996.

"Nylons (Polyamides)." MatWeb. http://www.matweb.com/reference/nylon.asp (accessed on October 24, 2005).

"Nylon Stockings." The Great Idea Finder. http://www.ideafinder.com/history/inventions/story062.htm (accessed on October 24, 2005).

"Polyamide 6 - Nylon 6 - PA 6." Azom.com. http://www.azom.com/details.asp?ArticleID=442 (accessed on October 24, 2005).

"Polyamides." Chemguide. http://www.chemguide.co.uk/organicprops/amides/polyamides.html (accessed on October 24, 2005).

$$\underset{\text{HO}}{\overset{\text{O}}{\underset{\|}{\text{C}}}} - \underset{\text{OH}}{\overset{\text{O}}{\underset{\|}{\text{C}}}}$$

OTHER NAMES:
Ethanedioic acid

FORMULA:
HOOCCOOH

ELEMENTS:
Carbon, hydrogen,
oxygen

COMPOUND TYPE:
Dicarboxylic acid;
organic acid (organic)

STATE:
Solid

MOLECULAR WEIGHT:
90.04 g/mol

MELTING POINT:
189.5°C (373.1°F);
begins to sublime at
157°C (315°F)

BOILING POINT:
Not applicable

SOLUBILITY:
Soluble in water and
ethyl alcohol; slightly
soluble in ether;
insoluble in most
other organic
solvents

K E Y   F A C T S

Oxalic Acid

OVERVIEW

Oxalic acid (ok-SAL-ik AS-id) is a transparent, colorless, crystalline solid that often occurs as the dihydrate (HOOC-COOH·2H$_2$O). The dihydrate melts and begins to decompose at 101.5°C (214.7°F), forming the anhydrous acid. The compound is odorless, but has a characteristic tart, acidic taste. The acid should never be tasted, however, as it is very toxic.

Oxalic acid is one of the first organic acids to have been discovered and studied. It was first isolated by the German chemist Johann Christian Wiegleb (1732-1800) in 1769 and first synthesized by the Swedish chemist Karl Wilhelm Scheele (1742-1786) in 1776. Friedrich Wöhler's (1800-1882) synthesis of oxalic acid entirely from inorganic materials was a critical step in disproving the Vitalistic Theory of chemistry. The theory claimed that compounds found in living organisms could be produced only by the act of some supernatural being, and not by human actions.

Oxalic acid occurs naturally in a number of vegetable products, including spinach, rhubarb, tea, chocolate, oats,

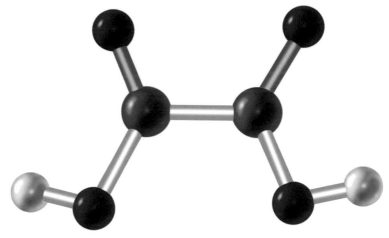

Oxalic acid. Red atoms are oxygen; white atoms are hydrogen; and black atoms are carbon. Gray sticks indicate double bonds. PUBLISHERS RESOURCE GROUP

pumpkin, lentils, beets, parsnips, and many kinds of nuts and berries. The amount present in foods is generally so low that it does not present risk to people who eat such products. Oxalic acid also occurs as a product of carbohydrate metabolism in animals.

HOW IT IS MADE

Traditionally, oxalic acid has been extracted from natural products by treating them with an alkaline solution, followed by crystallization of the acid. Sodium hydroxide is the alkaline material most commonly used for this procedure. Today, a number of methods are available for the commercial preparation of oxalic acid. In one procedure, carbon monoxide gas is bubbled through a concentrated solution of sodium hydroxide to produce oxalic acid. Alternatively, sodium formate (COONa) is heated in the presence of sodium hydroxide or sodium carbonate to obtain the acid. Another popular method of preparing oxalic acid involves the oxidation of sucrose (common table sugar) or more complex carbohydrates using nitric acid as a catalyst. The reaction results in the formation of oxalic acid and water as the primary products.

COMMON USES AND POTENTIAL HAZARDS

Over a half million kilograms (about 1 million pounds) of oxalic acid are produced in the United States each year.

Interesting Facts

- One family of plants, the Oxalidaceae, get their name from the high concentrations of oxalic acid they contain. Perhaps the best known member of the family is the wood sorrel.

- A number of molds produce oxalic acid as a major metabolic product. Some species of *Penicillium* and *Aspergillus*, for example, convert glucose into oxalic acid.

The greatest portion of the product is used in a variety of cleaning products, including substances for the bleaching and cleaning of wood, cork, cane, feathers, and natural and synthetic fibers. Many metal polishes, auto radiator cleaners, and laundry rinses also contain oxalic acid. Some rust-proofing materials also contain the compound.

The list of other household and industrial applications of oxalic acid is extensive and includes:

- In the printing and dyeing of fabrics, especially calico;

- For the treatment of leather products, especially the leather used to make book covers;

- As a paint and varnish remover as well as a stain remover for ink and rust marks;

- In the manufacture of blue ink, celluloid, rubber, and other synthetic products;

- For the decolorization of glycerol and the stabilization of hydrocyanic acid (HCN);

- As a purification agent for various chemicals, especially methanol (methyl alcohol); and

- In the preparation of certain medicines and pharmaceuticals.

Oxalic acid is a strong skin, eye, and respiratory irritant in pure form. It is also toxic by ingestion, causing nausea, vomiting, diarrhea, kidney damage, convulsions, coma, and

Words to Know

ANHYDROUS COMPOUND A compound that lacks any water of hydration.

HYDRATE A chemical compound formed when one or more molecules of water is added physically to the molecule of some other substance.

CATALYST A material that increases the rate of a chemical reaction without undergoing any change in its own chemical structure.

METABOLISM Process that includes all of the chemical reactions that occur in cells by which fats, carbohydrates, and other compounds are broken down to produce energy and the compounds needed to build new cells and tissues.

SUBLIME To go from solid to gaseous form without passing through a liquid phase.

SYNTHESIS A chemical reaction in which some desired chemical product is made from simple beginning chemicals, or reactants.

death. Since people are exposed to pure oxalic acid only in the workplace, these hazards are usually not of concern to most individuals. Of more concern to the general public is the possibility of ingesting unusually large amounts of foods containing oxalic acid. In the body, the acid tends to react with calcium ions (Ca^{2+}) forming calcium oxalate, which is insoluble. This reaction has two harmful effects. First, it removes calcium needed by the body for other biological functions, resulting in a calcium deficiency problem. Second, it may result in the precipitation of calcium oxalate crystals on the inner lining of blood vessels and the small tubes in the kidneys, reducing the flow of blood and resulting in the development of kidney disorders.

FOR FURTHER INFORMATION

"Material Safety Data Sheet." Hill Brothers Chemical Company. http://hillbrothers.com/msds/pdf/oxalic-acid.pdf (accessed on October 23, 2005).

"Oxalic Acid." Al's Home Improvement Center. http://alsnetbiz.com/homeimprovement/oxalic_acid.html (accessed on October 23, 2005).

"Oxalic Acid Dihydrate." Chemical Land 21. http://www.chemicalland21.com/arokorhi/industrialchem/ organic/OXALIC%20ACID.htm (accessed on October 23, 2005).

"The Rhubarb Compendium." Rhubarb Poison Information Center. http://www.rhubarbinfo.com/rhubarb-poison.html (accessed on October 23, 2005).

KEY FACTS

FORMULA:
Not applicable

ELEMENTS:
Carbon, hydrogen, oxygen, and other elements

COMPOUND TYPE:
Not applicable

STATE:
Solid

MOLECULAR WEIGHT:
Varies widely: 20,000 to 400,000 g/mol

MELTING POINT:
Not applicable

BOILING POINT:
Not applicable

SOLUBILITY:
Soluble in water; insoluble in organic solvents

Pectin

OVERVIEW

Pectin (PEK-tin) is a mixture, not a compound. Mixtures differ from compounds in a number of important ways. The parts making up a mixture are not chemically combined with each other, as they are in a compound. Also, mixtures have no definite composition, but consist of varying amounts of the substances from which they are formed.

Chemically, pectin is a polysaccharide, a very large molecule made of many thousands of monosaccharide units joined to each other in long, complex chains. Monosaccharides are simple sugars. The most familiar monosaccharide is probably glucose, the sugar from which the human body obtains the energy it needs to grow and stay healthy. The monosaccharides in pectin are different from and more complex than glucose.

Pectin occurs naturally in many fruits and vegetables. It is most abundant in citrus fruits such as lemons, oranges, and grapefruits, which may consist of up to 30 percent pectin. In pure form it is a yellowish-white powder with virtually no odor and a slightly gummy taste. When dissolved in water, it forms a thick, jelly-like mass. This property explains one of its primary purposes: the jelling of fruits when they are made into jams and jellies.

HOW IT IS MADE

Pectin is made naturally in ripening fruit. It is obtained commercially by treating the raw material (citrus peel or apple pomace) with hot, acidified water. (Apple pomace is the residue remaining after pressing of apples.) The pectin in the peel or apple pomace dissolves in the hot water and is then purified by repeated filtrations. It is extracted from the water solution by adding alcohol or an aluminum salt to the solution, causing the pectin to precipitate out of solution. The precipitate is then dried and ground into a powder.

Additional steps are sometimes carried out to convert the pectin produced by this method, called high ester pectin, to a form that is more soluble: low ester pectin. To achieve this change, high ester pectin is treated with either acids or alkalis, washed, and purified.

COMMON USES AND POTENTIAL HAZARDS

Pectin is used primarily as a jelling agent in the manufacture of jams and jellies. It also has a number of other applications as a food additive. For example, it is added to some yogurts to provide the consistency that allows the yogurt to hold its shape and still be capable of being stirred. It is added to concentrated fruit drinks to keep the solid and liquid components of the drink in suspension with each other. It is also an ingredient in fruit and milk desserts, added to ensure that the final product has the proper consistency and stability.

Pectin is also used as an additive in pharmaceutical and cosmetic preparations. It acts as an emulsifying agent, to stabilize the product, or to give it the proper consistency. In combination with an antibiotic, pectin has also been used as an anti-diarrheal agent. Some studies have shown that daily doses of pectin may have a small but significant lowering effect on cholesterol levels.

The U.S. Food and Drug Administration has classified pectin as an approved food additive. It is considered safe

Interesting Facts

- The role of pectin in the formation of jams and jellies from fruits was first recognized in the 1820s.

- The first commercial manufacture of pectin took place in Germany in 1908. Producers of apple juice found that they could use apple pomace, previously a waste product, to make a useful product that could be sold, pectin.

- The first recipes for the use of pectin to make jams and jellies date to the first century when the Roman writer Marcus Gavius Apicius wrote a recipe book "Of Culinary Methods.";

- American inventor Paul Welch was granted a patent in 1917 for the production of grape jam, using pectin. Welch called his product Grapelade and sold his entire production to the U.S. Army. The Army sent the Grapelade to troops serving in Europe in World War I (1914–1918). After the war, returning troops demanded more Grapelade, making it a popular consumer item in the United States.

for human consumption when used in normal amounts as a food additive. It may cause some digestive problems for people with allergies to citrus fruits. Some studies suggest that pectin may also inhibit the absorption of minerals such as zinc, copper, iron, and calcium, although this effect is not serious enough to prevent its use as a food additive.

Words to Know

EMULSION A temporary mixture of two liquids that normally do not dissolve in each other.

POLYSACCHARIDE A very large molecule made of many thousands of simple sugar molecules joined to each other in long, complex chains.

FOR FURTHER INFORMATION

"Genu® Pectin." CP Kelco.
http://www.cpkelco.com/food/pectin.html (accessed on December 22, 2005).

Knox, J. Paul, and Graham B. Seymour, eds. *Pectins and Their Manipulation.* Boca Raton, FL: CFC Press, June 2002.

"Pectin."
http://www.cfs.purdue.edu/class/f&n630/Virt_Class_2/pectin.htm (accessed on December 22, 2005).

"Pectin." PDRHealth.
http://www.pdrhealth.com/drug_info/nmdrugprofiles/nutsupdrugs/pec_0198.shtml (accessed on December 22, 2005).

See Also Gelatin

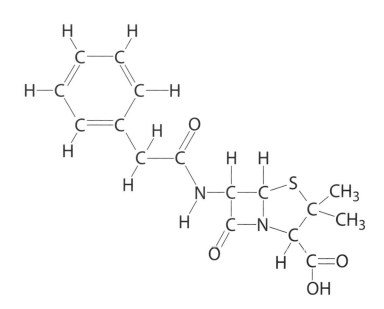

OTHER NAMES:
Not applicable; see
Overview

FORMULA:
$(CH_3)_2C_5H_3NSO$
$(COOH)NHCOR$, where
R represents any one
of a number of
substituted groups;
see Overview

ELEMENTS:
Carbon, hydrogen,
oxygen, nitrogen,
sulfur

COMPOUND TYPE:
Bicyclic acid
(organic)

STATE:
Solid

MOLECULAR WEIGHT:
Varies; see Overview
g/mol

MELTING POINT:
Varies; see Overview

BOILING POINT:
Not applicable; all
forms decompose
when heated above
their melting points

SOLUBILITY:
Slightly soluble in
water; soluble in
ethyl alcohol, ether,
chloroform and most
organic solvents

KEY FACTS

Penicillin

OVERVIEW

The penicillins (pen-uh-SILL-ins) are a class of antibiotic compounds derived from the molds *Penicillium notatum* and *Penicillium chrysogenum*. The class contains a number of compounds with the same basic bicyclic structure to which are attached different side chains. That basic structure consists of two amino acids, cysteine and valine, joined to each other to make a bicyclic ("two-ring") compound. The different forms of penicillin are distinguished from each other by adding a single capital letter to their names. Thus: penicillin F, penicillin G, penicillin K, penicillin N, penicillin O, penicillin S, penicillin V, and penicillin X. A number of other antibiotics, including ampicillin, amoxicillin, and methicillin, have similar chemical structures.

Penicillin was discovered accidentally in 1928 by the Scottish bacteriologist Alexander Fleming (1881-1995). Fleming noticed that a green mold, which he later identified as *Penicillium notatum*, had started to grow on a petri dish that he had coated with bacteria. As the bacteria grew towards the

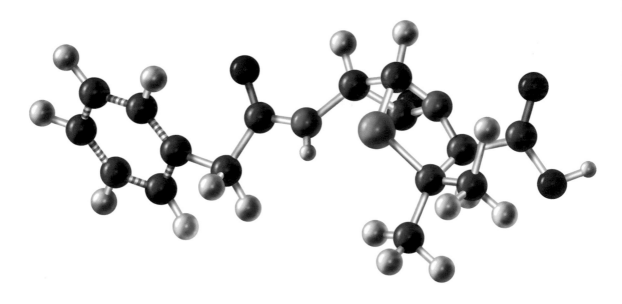

mold, they began to die. At first, Fleming saw some promise in this observation. Perhaps the mold could be used to kill the bacteria that cause human disease. His experiments showed, however, that the mold's potency declined after a short period of time He was also unable to isolate the anti-bacterial chemical produced by the mold. He decided that further research on Penicillium was probably not worthwhile.

As a result, it was not until a decade later that *Penicillium*'s promise was realized. In 1935, English pathologist Howard Florey (1898-1968) and his biochemist colleague Ernst Chain (1906-1970) came across Fleming's description of his experiment and decided to see if they could isolate the chemical product produced by *Penicillium* with anti-bacterial action. They were eventually successful, isolating and purifying a compound with anti-bacterial action, and, in 1941, began trials with human subjects to test its safety and efficacy (ability to kill bacteria). The successful conclusion of those trials not only provided one of the great breakthroughs in the human battle against infectious diseases, but also won for Florey, Chain, and Fleming the 1945 Nobel prize for Physiology or Medicine.

HOW IT IS MADE

Penicillins are classified as biosynthetic or semi-synthetic. Biosynthetic penicillin is natural penicillin. It is

Interesting Facts

- The discovery of penicillin's antibacterial action came at just the right time: the onset of World War II. The new drug made possible the saving of untold numbers of lives. As just one example, about three-quarters of all soldiers who developed bone infections as a result of wounds suffered in World War I died of those infections. By contrast, no more than about 5 percent of those who developed similar infections in World War II died. The availability of penicillin to treat those wounds in World War II made the difference in those survival rates.

produced by culturing molds in large vats and collecting and purifying the penicillins they produce naturally. There are six naturally occurring penicillins. The specific form of penicillin produced in a culturing vat depends on the nutrients provided to the molds. Of the six natural penicillins, only penicillin G (benzylpenicillin) is still used to any extent.

Semi-synthetic penicillins are produced by making chemical alterations in the structure of a naturally occurring penicillin. For example, penicillin V is made by replacing the $-CH_2C_6H_5$ group in natural penicillin G with a $-CH_2OC_6H_5$ group.

COMMON USES AND POTENTIAL HAZARDS

Penicillins are prescription medications used to treat a variety of bacterial infections, including meningitis, syphilis, sore throats, and ear aches. They do so by inactivating an enzyme used in the formation of bacterial cell walls. With the enzyme inactivated, bacteria can not make cell walls and die off. Penicillins do not act on viruses in the same way they do on bacteria, so they are not effective against viral diseases, such as the flu or the common cold.

A number of side effects are related to the use of penicillin. These side effects include diarrhea, upset stomach, and vaginal yeast infections. In those individuals who are allergic

Words to Know

BICYCLIC Refers to a molecule in which the atoms are arranged so as to look like two rings.

BIOSYNTHETIC Natural; made from a living organism.

SEMI-SYNTHETIC Produced by making chemical alterations in the structure of a naturally occurring organism.

to penicillins, side effects are far more serious and include rash, hives, swelling of tissues, breathing problems, and anaphylactic shock, a life-threatening condition that requires immediate medical treatment.

Penicillin may alter the results of some medical tests, such as those for the presence of sugar in the urine. Penicillin can also interact with a number of other medications, including blood thinners, thyroid drugs, blood pressure drugs, birth control pills, and other antibiotics, in some cases decreasing their effectiveness.

Once promoted as wonder drugs, the use of penicillins has declined slowly because of the spread of antibiotic resistance. Antibiotic resistance occurs when new strains of bacteria evolve that are resistant to existing types of penicillin. One reason that antibiotic resistance has become a problem is the extensive and often unnecessary use of penicillins. When they are prescribed for colds and the flu, for example, they have no effect on the viruses that cause those diseases, but they encourage the growth of bacteria more able to survive against penicillins.

FOR FURTHER INFORMATION

"β-Lactam Antibiotics: Penicillins." *The Merck Manual of Diagnosis and Therapy.* Chapter 153. Available online at http://www.merck.com/mrkshared/mmanual/section13/chapter153/153b.jsp (accessed on October 23, 2005).

Chain, E. B. "The Chemical Structure of the Penicillins." Nobel Lecture, March 20, 1945. Available online at http://nobelprize.org/medicine/laureates/1945/chain-lecture.pdf (accessed on October 23, 2005).

Moore, Greogry A., and Ollie Nygren. "Penicillins." The Nordic Expert Group for Criteria Documentation of Health Risks from Chemicals. http://ebib.arbetslivsinstitutet.se/ah/2004/ah2004_06.pdf (accessed on October 23, 2005).

Ross-Flanigan, Nancy. "Penicillins." *Gale Encyclopedia of Medicine.* Edited by Jacqueline L. Longe and Deidre S. Blanchfield. 2nd ed. Detroit, MI: Gale, 2002.

"Tom Volk's Fungus of the Month for November 2003." http://botit.botany.wisc.edu/toms_fungi/nov2003.html (accessed on October 23, 2005).

$$\underset{\overset{|}{\underset{-O}{}}}{\overset{\overset{O}{\parallel}}{Cl}}\overset{O}{\underset{O}{}}$$

OTHER NAMES:
See Overview

FORMULA:
$-ClO_4$

ELEMENTS:
Chlorine and oxygen, in combination with other ions

COMPOUND TYPE:
Inorganic salts

STATE:
Solid

MOLECULAR WEIGHT:
117.49 to 138.55 g/mol

MELTING POINT:
Ammonium perchlorate: Decomposes explosively when heated; Potassium perchlorate: 525°C (977°F); Sodium perchlorate: 480°C (896°F)

BOILING POINT:
Not applicable; all decompose at or above melting points

SOLUBILITY:
Soluble in water; ammonium perchlorate is also soluble in methyl alcohol and slightly soluble in ethyl alcohol

KEY FACTS

Perchlorates

OVERVIEW

The perchlorates (per-KLOR-ates) are a family of compounds consisting of salts of perchloric acid, $HClO_4$. The family consists of dozens of compounds, the most important of which are ammonium perchlorate (NH_4ClO_4), potassium perchlorate ($KClO_4$), and sodium perchlorate ($NaClO_4$).

Although perchlorates have been known since the early nineteenth century, they were not produced commercially until the 1890s. Even then, they were not produced in large volumes until World War II (1939–1945), when they were made for use in explosives. Perchlorate production continued to grow in the post-war period, especially during the military build-up that accompanied the Cold War (1945–1991) between the United States and the Soviet Union. In recent years, however, concerns about the presence of perchlorates in water supplies have become increasingly widespread.

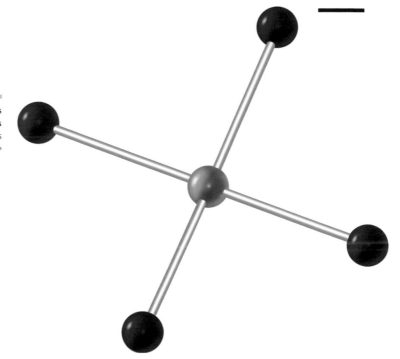

Perchlorate ion. Red atoms are oxygen. Green atom is chlorine. PUBLISHERS RESOURCE GROUP

HOW IT IS MADE

Each perchlorate is produced by a different method. Ammonium perchlorate, for example, is produced in a reaction among ammonium hydroxide (NH_4OH), hydrochloric acid (HCl), and sodium chlorate ($NaClO_3$).

Potassium perchlorate can be made in two ways. One method by electrolyzing potassium chlorate ($KClO_3$) in water. The other is by heating potassium chlorate ($KClO_3$). This results in a mixture of potassium perchlorate and potassium chloride (KCl), which is removed.

Sodium perchlorate is made by heating a mixture of sodium chlorate and sodium chloride. This reaction yields sodium chloride (NaCl) as well as the sodium perchlorate. The excess sodium chloride is removed from the reaction, leaving sodium perchlorate behind.

COMMON USES AND POTENTIAL HAZARDS

By far the most important uses of perchlorates are in the production of explosives and fireworks and as jet and rocket

Interesting Facts

- Small amounts of potassium perchlorate are found naturally in Chile in deposits of sodium nitrate.

- About 9 million kilograms (20 million pounds) of perchlorates are produced each year in the United States.

- In January 2005, the U.S. Environmental Protection Agency set a standard of 0.007 milligrams per kilogram of body weight as the maximum recommended dose a person should ingest of perchlorates per day.

- Between 1997 and 2004, Lockheed Martin spent $80 million to study, clean up, and replace local water systems that had been contaminated with perchlorates used by the company in the production of military weapons and rocket fuels. The two systems were located near Riverside and Redlands, California.

fuels. These uses are based on the fact that all perchlorates are very unstable oxidizing agents. An oxidizing agent is a material that supplies oxygen to some other substances or removes electrons from that substance. When used in explosives, the perchlorates supply oxygen to the fuels in the explosive (such as sulfur or carbon), causing them to burn very rapidly, producing large volumes of gases in a very short time. When used as a rocket or jet fuel, perchlorates supply the oxygen needed to burn the fuel itself, such as kerosene or hydrogen.

Some perchlorates have other, more limited, uses. For example, potassium perchlorate was previously used to treat Graves disease, a condition in which the body produces too much thyroid hormone. It is still used to monitor the production of thyroid hormones. Potassium perchlorate is also used in emergency breathing equipment for high altitude aircraft and underwater boats. Other uses of perchlorates include:

- In nuclear reactors and electronic tubes;

- As additives in lubricating oils;

- In the tanning and finishing of leather products;

Words to Know

ELECTROLYSIS A process in which an electric current is used to bring about chemical changes.

SALT An ionic compound where the anion is derived from an acid.

SYNTHESIS A chemical reaction in which some desired chemical product is made from simple beginning chemicals, or reactants.

- As a fixer for fabrics and dyes;
- In electroplating;
- In the refining of aluminum metal;
- In the manufacture of rubber products;
- In the production of certain paints and enamels.

Perchlorates pose a serious risk to humans because they are unstable and have a tendency to explode spontaneously. They are also human health hazards, with harmful effects on both the brain and the thyroid. They have a tendency to prevent the uptake of iodine by the thyroid, thus interfering with the synthesis of hormones normally produced by that organ. Health officials believe that exposure to perchlorates may cause infertility in women or may have harmful effects on their newborn children. These effects include mental retardation and delays in their normal development. As a consequence, efforts are underway to locate areas where perchlorates may have entered the public water supply causing potential health problems for people living in the region.

FOR FURTHER INFORMATION

Perchlorate News.com.
http://www.perchloratenews.com/index.html (accessed on December 10, 2005).

"Perchlorates: New Report on Widespread Rocket Fuel Pollution in Nation's Food and Water." Organic Consumers Association. http://www.organicconsumers.org/perchlorate.htm (accessed on December 10, 2005).

"Potassium Perchlorate." J. T. Baker. http://www.jtbaker.com/msds/englishhtml/P5983.htm (accessed on December 10, 2005).

"Safety (MSDS) Data for Ammonium Perchlorate." Physical and Theoretical Chemistry Lab. http://ptcl.chem.ox.ac.uk/MSDS/AM/ammonium_perchlorate.html (accessed on December 10, 2005).

Sharp, Renee, and Bill Walker. *Rocket Science: Perchlorate and the Toxic Legacy of the Cold War.* Washington, D.C.: Environmental Working Group, July 2001. Also available online at http://www.ewg.org/reports_content/rocketscience/perchlorate.pdf (accessed on December 10, 2005).

"Toxicological Profile for Perchlorates." Agency for Toxic Substances and Disease Registry. http://www.atsdr.cdc.gov/toxprofiles/tp162.html (accessed on December 10, 2005).

OTHER NAMES:
Petroleum jelly;
paraffin jelly;
vasoliment; liquid
paraffin; mineral
oil; paraffin oil

FORMULA:
Not applicable

ELEMENTS:
Carbon, hydrogen

COMPOUND TYPE:
Not applicable

STATE:
Semi-solid or liquid

MOLECULE WEIGHT:
Not applicable

MELTING POINT:
Not applicable

BOILING POINT:
Not applicable

SOLUBILITY:
Insoluble in water and
ethyl alcohol; soluble
in benzene,
chloroform, ether,
carbon disulfide, and
other organic
solvents

K E Y   F A C T S

Petrolatum

OVERVIEW

Petrolatum (peh-tro-LAY-tum) is a mixture, not a compound. Mixtures differ from compounds in a number of important ways. The parts making up a mixture are not chemically combined with each other, as they are in a compound. Also, mixtures have no definite composition, but consist of varying amounts of the substances from which they are formed.

Petrolatum is a complex mixture of hydrocarbons derived from the distillation of petroleum. Hydrocarbons are compounds that contain only carbon and hydrogen. The hydrocarbons that make up petrolatum belong to the methane (saturated or alkane) family of hydrocarbons with the general formula C_nH_{2N+2}. Some members of the family include methane (CH_4), ethane (C_2H_5), propane (C_3H_8), and butane (C_4H_{10}).

Petrolatum occurs in a semi-solid or liquid form. The semi-solid form is also called petroleum jelly or mineral jelly and is commercially available under a number of trade names, including Kremoline, Pureline, Sherolatum, and Vaseline™. It ranges in color from white to yellowish to amber. It is practically odorless and tasteless. It melts over a wide range, from about 38°C to about 55°C (100°F to 131°F). The liquid form is also known as liquid paraffin, mineral oil, or white mineral oil. Such products are sold commercially under trade names such as Alboline, Drakeol, Frigol, Kremol, and Paroleine. It is a colorless, tasteless, and odorless oily liquid.

Oil was first discovered in the United States in the 1850s in western Pennsylvania. A chemist from Brooklyn, New York, Robert Augustus Chesebrough (1837-1938), visited the new wells and noticed a wax-like material sticking to the petroleum drilling rods. He learned that oil workers used the "rod wax" to heal burns on their skin. Chesebrough eventually extracted and purified the substance—petrolatum—from petroleum and began manufacturing it in 1870. He received several patents for his discovery and in 1878, he gave his product the trade name of Vaseline™. His product quickly became popular as an ointment for wounds and burns. Unlike the animal and vegetable oils then being used for that purpose, petrolatum did not spoil. By the late 1870s, Vaseline™ was selling at the rate of one jar every minute in the United States. In 1880, it was added to the U.S. Pharmacopoeia, a manual that lists drugs used in medical practice.

HOW IT IS MADE

Petrolatum is a product of the fractional distillation of crude oil. Crude oil is a complex mixture of hundreds or thousands of compounds. These compounds can be separated, or distilled, from each other by heating crude oil to high temperatures. As the temperature of the crude oil rises, various groups or a "fraction" of compounds boil off. The first group of compounds includes gaseous compounds dissolved in crude oil. The next group of compounds includes compounds with low boiling points. The next group of compounds includes compounds with slightly higher boiling points. And so on. Eventually, a tar-like mass of compounds with very high boiling points is left behind in the distilling tower. This residue is heated to separate liquids from solids remaining behind. Some of these liquids and solids make up the semi-solid and liquid forms of petrolatum.

Interesting Facts

- Both solid and liquid petrolatum are available in three grades, known as USP (U.S. Pharmacopoeia), NF (National Formulary), and FCC (Food Chemicals Codex).

- A synthetic version of petrolatum is made from soybean oil as an alterna-tive to petroleum-based petrolatum. It is used primarily in the manu-facture of cosmetics.

- Skin care products gener-ally contain petrolatum in a concentration of about 1 to 3 percent.

COMMON USES AND POTENTIAL HAZARDS

Petrolatum has a wide variety of uses, ranging from personal care and medical applications to industrial uses. The solid form, such as Vaseline^{TM} is used as a topical oint-ment for the treatment of dry, cracked skin and to reduce the risk of infection. It works as a moisturizing agent because it reduces water loss from the skin. It helps prevent infection because it creates a barrier over wounds that prevents dis-ease-causing organisms from entering the body. Solid petro-latum is also an ingredient in many skin care and cosmetic products, such as skin lotions, body and facial cleansers, anti-perspirants, lipsticks, lip balms, sunscreens, and after-sun lotions. In hair products, it helps smooth frizzy hair by allowing hair to retain its natural moisture. The formation used in most of these products remains virtually unchanged from that developed by Robert Chesebrough in the 1800s.

Solid petrolatum is also used in industrial applications for a variety of purposes, such as:

- As a softener in the production of rubber products;

- In the food processing industry, to coat raw fruits and vegetables and to help products retain moisture;

- As a defoaming agent in the production of beet sugar and yeasts;

- For the lubrication of firearms and machine parts;

Words to Know

FRACTIONAL DISTILLATION The process of extracting compounds from petroleum by heating the petroleum and collecting the individual compounds as they boil off when their boiling points are reached.

HYDROCARBON A compound that contains hydrogen and carbon atoms.

MIXTURE A collection of two or more elements and/or compounds with no definite composition.

- In the production of modeling clays;
- In the manufacture of candles, to prevent a candle from shrinking as it cools after being burned;
- In the preparation of shoe polishes; and
- As an ingredient in rust preventatives.

The primary use of liquid petrolatum is as a laxative, a product that loosens the bowels. It also has a number of other applications, such as an additive in foods such as candies, confectionary products, and baked goods; as an ingredient in personal care products, such as baby oil creams, hair conditioning lotions, and ointments; in many different kinds of pharmaceutical preparations; in the production of industrial lubricants; as a softening agent in the manufacture of rubber, textiles, fibers, adhesives, and machine parts; as dust suppressants; and as dehydrating agents for a number of industrial processes.

FOR FURTHER INFORMATION

"Another Old-Fashioned Product Vindicates Itself." *Medical Update* (October 1992): 6.

Morrison, David S. "Petrolatum: A Useful Classic." *Cosmetics and Toiletries* (January 1996): 59-68. Also available online at http://www.penreco.com/newsevents/tradearticles/petrolatum classic.pdf (accessed on December 22, 2005).

"Material Safety Data Sheet." Penreco. http://www.penreco.com/products/pdfs/petrolatum/ pen00421technicalpetrolatum.pdf (accessed on December 22, 2005).

Penreco (Petrolatum company). http://www.penreco.com/index.asp (accessed on December 22, 2005).

Schramm, Daniel. "The North American USP Petrolatum Industry." *Soap & Cosmetics* (January 2002): 60-63.

See Also Petroleum

KEY FACTS

Petroleum

OVERVIEW

Petroleum (peh-TRO-lee-yum) is a mixture, not a compound. Mixtures differ from compounds in a number of important ways. The parts that make up a mixture are not chemically combined with each other, as they are in a compound. Also, mixtures have no definite composition, but consist of varying amounts of the substances of which they are formed.

Petroleum is a very complex mixture of hydrocarbons—compounds that consist of carbon and hydrogen only. Small amounts of other organic compounds containing oxygen, sulfur, nitrogen, phosphorus, and other elements are also present. The hydrocarbons that make up petroleum include the alkanes, alkenes, alkynes, cyclic hydrocarbons, and aromatic hydrocarbons. Alkanes are hydrocarbons in which carbon and hydrogen are bonded to each other by single bonds. In alkenes, at least one double bond is present. In alkynes, carbon and hydrogen atoms are bonded by at least one triple bond. In cyclic hydrocarbons, the carbon atoms are arranged in a ring. In aromatic hydrocarbons, the carbon atoms are joined to each other in a ring that has alternate single and double bonds. Many of the hydrocarbons in petroleum also have side chains that also consist of carbon and hydrogen atoms.

The physical properties of petroleum vary somewhat, depending on the source from which it comes and its composition. Generally, it is a thick, oily liquid that is dark yellow to brown to greenish-black in color, with a strong, unpleasant odor.

HOW IT IS MADE

Scientists believe that petroleum was formed about 300 million years ago. When microscopic plants and animals that lived in the ocean died and sank to the bottom, they were gradually covered and compressed with more layers of organic material along with sand and mud. As the mud grew thicker, it created pressure on the organic material, causing it to become increasingly warmer. The heat and pressure caused the organic matter to decay in the absence of oxygen, converting it into petroleum and natural gas.

Over time, the primitive oceans dried up. The sand and mud that had accumulated on the ocean floors changed into rock. The natural gas and liquid petroleum that had formed on the ocean floor was trapped in the rock. It flowed through cracks in the rock until it reached porous rock that acted like a sponge and soaked up the petroleum and natural gas. These fossil fuels remain trapped in the porous rock by non-porous layers of rock that act like caps or seals on the porous rocks.

Geologists have discovered a number of ways of finding these oil- and gas-soaked reservoirs. They measure changes in the Earth's magnetic field and use electronic "sniffers" that can detect hydrocarbons trapped in rock. Once oil is found, prospectors extract it by drilling into the porous rock. The pressure of gases within the porous rock pushes petroleum up to the surface, where it can be captured and stored.

Liquid petroleum has no commercial value as it comes from the earth. After it is captured from wells, therefore, it has to be refined, or separated into useful components. Refining is a process by which petroleum is heated to high temperatures in a tall cylindrical tower. The many different hydrocarbons

Interesting Facts

- The United States uses by far more petroleum than any other nation in the world. It uses about 1,000 billion liters (250 billion gallons) every year. About half that amount is imported from other countries, primarily in the Middle East.

- The average fuel economy of automobiles produced in the United States has decreased steadily since 1985.

in petroleum boil off and rise upward in the tower. The higher they rise, the more they cool off.

Each hydrocarbon eventually reaches a height at which it changes back to a liquid. Traps are inserted at various heights in the tower to catch each hydrocarbon as it changes back to a liquid. One level of traps, for example, catches hydrocarbons that condense to a liquid between about 40°C and 170°C (100°F and 340°F). This "fraction" is called the *gasoline* or *petrol* fraction of petroleum. Another level of traps catches hydrocarbons that condense between temperatures of about 170°C and 250°C (340°F and 480°F). This group of hydrocarbons is called the *kerosene* fraction. The other major fractions obtained from petroleum are the diesel and fuel oil fractions. Anything that boils above about 400°C (750°F) is referred to as the residual oil fraction. In many cases, each of the fractions obtained by this process can be further refined to separate it into even smaller fractions.

COMMON USES AND POTENTIAL HAZARDS

By far the most common use of petroleum is as gasoline for automobiles and trucks. About 90 percent of the petroleum used in the United States fuels vehicles powered by internal combustion engines. Diesel fuel is used to power trucks, buses, trains, and some automobiles. Heating oil warms buildings and powers industrial boilers. Utilities use residual fuel oil to generate electricity. Jet fuel, produced from kerosene, powers airplanes. Some airplanes use aviation gasoline, which has a higher octane number than automobile gasoline. Octane number is a measure of a fuel's efficiency.

Kerosene is used to power lamps and heaters. Power plants and factories use petroleum coke, a solid fuel made of petroleum, as a fuel. Asphalt and tar, solid materials left over after the fractionation of petroleum, are components of paved roads. Petrolatum, another solid component of petroleum, is used as a lubricant and moisturizer. Paraffin wax, also obtained as a by-product of petroleum distillation, is an ingredient in candles, candy, polishes, and matches. Other forms of petroleum are used as lubricating oils for engines or as solvents in paints.

A second very important use of petroleum is in the manufacture of plastics and other chemicals. The number of chemical compounds obtained from petroleum and used as raw materials in chemical reactions is almost endless. It includes compounds such as methane, ethane, propane, butane, ethene (ethylene), propene (propylene), butene (butylene), benzene, methanol (methyl alcohol), ethanol (ethyl alcohol), phenol, xylene, naphthalene, and anthracene, to mention only a few.

There are many drawbacks associated with using petroleum as a fuel. One of the most important is environmental pollution. Burning petroleum products releases greenhouse gases—gases that tend to accumulate in the atmosphere and contribute to a gradual warming of the Earth's annual average temperature. Other pollutants produced when petroleum products are burned include sulfur oxides and oxides of nitrogen, compounds that contribute to the development of acid rain, and carbon monoxide and ozone, pollutants that damage plant and animal life as well as physical structures, like buildings and statues.

The production and transportation of petroleum products also poses environmental problems. Oil spills that occur during any phase of petroleum production and use may kill aquatic life and pollute an area for years. The worst oil spill in North America, for example, occurred during the wreck of the oil tanker *Exxon Valdez* in 1989, when 40,000 tons of crude oil were dumped into Alaska's Prince William Sound. The worst oil spill in the world occurred in 1978 as a result of the wreck of the oil tanker *Amoco Cadiz* off the coast of Brittany, France. More than 220,000 tons of oil were released during that accident.

Words to Know

DISTILLATION A process of separating two or more substances by boiling the mixture of which they are composed and condensing the vapors produced at different temperatures.

HYDROCARBON A compound that contains hydrogen and carbon atoms.

MISCIBLE Able to be mixed; especially applies to the mixing of one liquid with another.

MIXTURE A collection of two or more elements and/or compounds with no definite composition.

FOR FURTHER INFORMATION

American Petroleum Institute. http://api-ec.api.org/newsplashpage/index.cfm (accessed on December 22, 2005).

Hyne, Norman J. *Nontechnical Guide to Petroleum Geology, Exploration, Drilling and Production,* 2nd ed. Tulsa, OK: Pennwell Books, 2001.

Leffler, William L. *Petroleum Refining in Nontechnical Language.* Tulsa, OK: Pennwell Books, 2000.

"Oil and Gas-Energy for the World." Institute of Petroleum. http://www.energyinst.org.uk/education/oilandgas/energy.htm (accessed on December 22, 2005).

"Petroleum." U.S. Energy Information Administration. http://www.eia.doe.gov/oil_gas/petroleum/info_glance/petroleum.html (accessed on December 22, 2005).

"Petroleum Refining Processes." OSHA Technical Manual. Washington, DC: Occupational Safety and Health Administration, 1999.. Available online at http://www.osha.gov/dts/osta/otm/otm_iv/otm_iv_2.html#1 (accessed on December 22, 2005).

See Also Methane; Petrolatum; Propane

OTHER NAMES:
Hydroxybenzene;
carbolic acid; phenylic
acid; benzophenol;
phenic acid

FORMULA:
C_6H_5OH

ELEMENTS:
Carbon, hydrogen,
oxygen

COMPOUND TYPE:
Aromatic alcohol
(organic)

STATE:
Solid

MOLECULAR WEIGHT:
94.11 g/mol

MELTING POINT:
40.89°C (105.6°F)

BOILING POINT:
181.87°C (359.37°F)

SOLUBILITY:
Soluble in water,
ethyl alcohol, ether,
chloroform, acetone,
benzene, and other
organic solvents

KEY FACTS

Phenol

OVERVIEW

Phenol (FEE-nol) is a white, crystalline solid with a characteristic odor and a sharp, burning taste. It tends to turn pink or pale red when exposed to light if not perfectly pure. Phenol has a tendency to absorb moisture from the air, changing into an aqueous solution of the compound. Such solutions have a slightly sweet flavor.

Phenol was probably first observed by German chemist Johann Rudolf Glauber (1604-1668). Glauber obtained phenol by condensing coal tar vapors and separating them into individual compounds. Coal tar is a thick black liquid left over when coal is heated in the absence of air to make coke. Phenol was largely ignored for almost 200 years until another German chemist, Friedlieb Ferdinand Runge (1795-1867), isolated phenol in 1834, also from coal tar. Runge gave the name *carbolic acid* to his discovery, a name that is still used occasionally for the compound. In 1843, French chemist Charles Frederic Gerhardt (1816-1856) suggested the modern name of *phenol* for the compound.

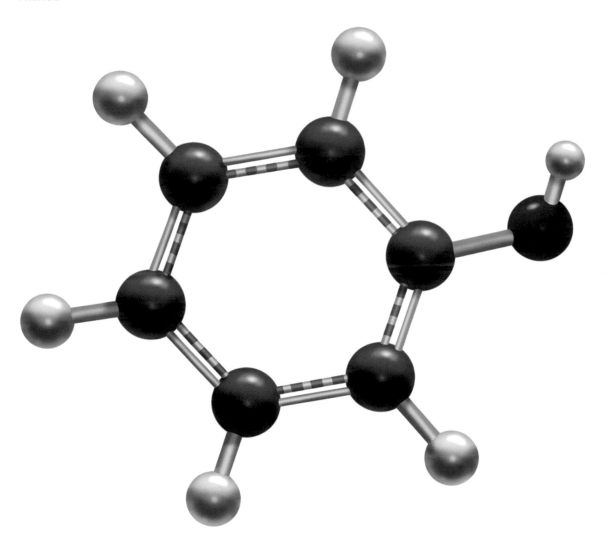

Phenol. Black atoms are carbon; red atom is oxygen; white atoms are hydrogen. Bonds in the benzene ring are represented by the double striped sticks. White sticks show single bonds. PUBLISHERS RESOURCE GROUP

The term *phenol* is also used to describe a class of aromatic organic compounds. Such compounds have a six-carbon ring structure like that of benzene (C_6H_6) to which is attached one or more hydroxyl (-OH) groups. Benzene is the simplest member of this group. Other members include the cresols ($CH_3C_6H_4OH$), xylenols (($CH_3)_2C_6H_3OH$), and the resorcinols ($C_6H_4(OH)_2$). The production of phenol amounts to about 6 million metric tons (about 6.5 million short tons) annually in the United States. The largest single use of phenol

Interesting Facts

- The only source of phenol until World War I (1914–1918) was coal tar.

- German chemist Friedrich Raschig (1863-1928) developed the first synthetic method for making phenol in 1915. The process involves the hydrolysis of chlorobenzene (C_6H_5Cl).

- Demand for phenol increased significantly in 1909 when German chemist Leo Hendrik Baekeland (1863–1944) announced the discovery of a new synthetic material, Bakelite, made from phenol and formaldehyde (CH_2O). Bakelite was the first thermosetting plastic ever discovered. A thermosetting plastic is one that, once shaped into some form, cannot be remelted again.

- The use of carbolic acid as an antiseptic was first suggested by Sir Joseph Lister (1827–1912) in 1865. Lister found that sterilizing medical instruments and treating wounds with carbolic acid dramatically reduced the number of deaths resulting from post-operative infections.

- The effectiveness of a substance in preventing bacterial infection is still expressed by a measure known as the *phenol coefficient*, the effectiveness of the material compared to that of a comparable amount of phenol.

is in the production of bisphenol A (($CH_3)_2C(C_6H_4OH)_2$) and a variety of plastics, primarily formaldehyde resins.

HOW IT IS MADE

In the early twenty-first century, virtually all of the phenol made is produced from cumene ($C_6H_5CH(CH_3)_2$). Cumene is first oxidized to produce cumene hydroperoxide: $C_6H_5CH(CH_3)_2 + O_2 \rightarrow C_6H_5(COOH)(CH_3)_2$. The cumene hydroperoxide is then treated with concentrated sulfuric acid (H_2SO_4) to obtain a mixture of acetone (CH_3COCH_3) and phenol: $C_6H_5(COOH)(CH_3)_2 + H_2SO_4 \rightarrow CH_3COCH_3 + C_6H_5OH$.

COMMON USES AND POTENTIAL HAZARDS

The primary use for phenol is as an intermediary chemical, a compound used in the synthesis of other chemicals. About 40 percent of all the phenol produced in the United States is used to make bisphenol A, while a similar amount is used in the production of a variety of polymers, such as phenol-formaldehyde plastics and nylon-6. The third largest application of phenol is in the manufacture of a host of other chemicals, xylene and aniline being the most important.

Phenol is no longer widely used as an antiseptic, partly because more efficient substances have been developed and partly because phenol may cause irritation and burning of the skin after prolonged use. The compound is still used in low concentrations in a number of health and medical applications, however, as in antipruritics (substances that reduce or prevent itching), cauterizing agents (substances for the burning of tissue by heat or chemicals), topical anesthetics (anesthetics used on the skin), chemexfoliants (substances that remove skin), throat sprays and lozenges (such as Chloraseptic®, Ambesol®, and Cepastat®), and in skin ointments (such as PRID salve® and CamphoPhenique® lotion). Phenol is also used in combination with slaked lime (calcium hydroxide; $Ca(OH)_2$) and other materials as a disinfectant for toilets, stables, cesspools, floors, and drains.

Phenol is a highly corrosive material that can cause serious burns to the skin, eyes, and respiratory system. It is toxic if ingested. The compound can enter the body in a variety of ways, such as absorption through the skin, inhalation of vapors, and ingestion of the solid compound or its solutions. Symptoms of phenol poisoning include nausea, vomiting, headache, respiratory failure, muscular weakness, severe depression, collapse, coma, and death. Skin exposure may cause redness, blisters, and/or minor to severe chemical burns. Ingestion of phenol may cause damage to the central nervous system, lungs, kidneys, liver, pancreas, and spleen. The concentration of phenol in most consumer and industrial products is low enough not to cause health problems for users. However, prolonged exposure to such products or their overuse may result in serious health risks. Handling of the pure compound or concentrated solutions of phenol also involves health hazards.

Words to Know

AQUEOUS SOLUTION A solution that consists of some material dissolved in water.

HYDROLYSIS The process by which a compound reacts with water to form two new compounds.

POLYMER A compound consisting of very large molecules made of one or two small repeated units called monomers.

SYNTHESIS A chemical reaction in which some desired chemical product is made from simple beginning chemicals, or reactants.

FOR FURTHER INFORMATION

"Occupational Safety and Health Guideline for Phenol." Occupational Safety and Health Administration.
http://www.osha.gov/SLTC/healthguidelines/phenol/recognition.html (accessed on December 29, 2005).

"Phenol." Greener Industry.
http://www.uyseg.org/greener_industry/pages/phenol/1PhenolAnnualProd.htm (accessed on December 29, 2005).

"Phenol." International Programme on Chemical Safety.
http://www.inchem.org/documents/pims/chemical/pim412.htm (accessed on December 29, 2005).

"ToxFAQs™ for Phenol." Agency for Toxic Substances and Disease Registry.
http://www.atsdr.cdc.gov/tfacts115.html (accessed on December 29, 2005).

Tyman, J. H. P. *Synthetic and Natural Phenols.* Amsterdam: Elsevier, 1996.

See Also Cumene; Polyamide 6,6

$$HO - \underset{\underset{O}{\|}}{\overset{\overset{OH}{|}}{P}} - OH$$

OTHER NAMES:
Orthophosphoric acid

FORMULA:
H₃PO₄

ELEMENTS:
Hydrogen, phos-
phorus, oxygen

COMPOUND TYPE:
Inorganic acid

STATE:
Solid. See Overview

MOLECULAR WEIGHT:
98.00 g/mol

MELTING POINT:
42.4°C (108°F)

BOILING POINT:
407°C (765°F)

SOLUBILITY:
Very soluble in water
and ethyl alcohol

KEY FACTS

Phosphoric Acid

OVERVIEW

Phosphoric acid (fos-FOR-ik AS-id) melts at a temperature just above room temperature (about 20°C; 68°F), so would be expected to occur as a solid under those conditions. As a solid, the acid is a white crystalline material with a strong tendency to absorb moisture from the air. In fact, phosphoric acid may also occur as a supercooled liquid at room temperature. A supercooled liquid is one that remains in a liquid state at temperatures below its freezing point, at which temperature it would be expected to be a solid. As a liquid, phosphoric acid is a colorless, odorless, syrupy liquid whose character is sometimes described as sparkling.

Phosphoric acid was discovered independently as a component of bone ash in 1770 by two Swedish chemists, Johann Gottlieb Gahn (1745-1818) and Karl Wilhelm Scheele (1742-1786). Four years later, Scheele discovered that the acid could be made by adding nitric acid to phosphorus.

Phosphoric acid is the ninth highest volume chemical produced in the United States. In 2004, the U.S. chemical

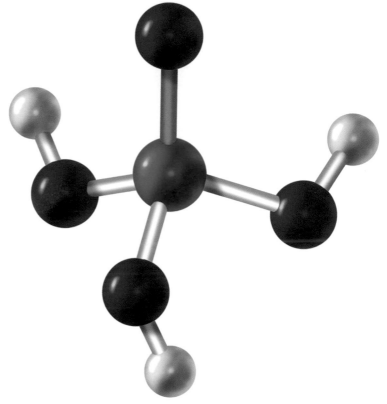

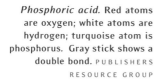

Phosphoric acid. Red atoms are oxygen; white atoms are hydrogen; turquoise atom is phosphorus. Gray stick shows a double bond. PUBLISHERS RESOURCE GROUP

industry made about 5.2 million kilograms (11.5 million pounds) of phosphoric acid. About 90 percent of that amount went to the manufacture of fertilizers.

HOW IT IS MADE

The most economical method for making phosphoric acid is by treating phosphate rock with sulfuric acid (H_2SO_4). Phosphate rock is naturally-occurring rock with large amounts of calcium phosphate ($Ca_3(PO_4)_2$). The product of this reaction is generally not very pure, but sufficiently pure for use in the production of fertilizers. Higher quality phosphoric acid can be made by burning phosphorus or phosphate rock in an electric furnace, converting either to gaseous phosphoric oxide (diphosphorus pentoxide; P_2O_5). The oxide is then dissolved in water to form the acid.

Interesting Facts

- Phosphoric acid is commercially available in a number of technical grades, ranging from agricultural (relatively low purity) to technical (from 50 to 100 percent purity) to FCC (Food Chemicals Codes) quality of at least 75 percent phosphoric acid.

COMMON USES AND POTENTIAL HAZARDS

By far the most important use of phosphoric acid is in the production of fertilizers. At least four major types of fertilizers are made from phosphoric acid: diammonium phosphate $((NH_4)_2HPO_4;$ DAP), monoammonium phosphate $(NH_4H_2PO_4;$ MAP), granulated triple superphosphate (GTSP), and superphosphoric acid, the only liquid among the group. An additional 5 percent of the phosphoric acid produced is used as an animal feed supplement.

The remaining 5 percent of phosphoric acid produced is used in a very wide range of commercial, industrial, and household products, including:

- For pickling cleaning and treating metallic surfaces, especially in the steel industry;
- In the synthesis of inorganic chemical compounds;
- As a catalyst in the manufacture of ethanol (ethyl alcohol), ethylene, and other organic compounds;
- As a food additive in a number of products, such as colas, beers, jams, and cheeses, where it adds a touch of tartness to the product;
- In the dentistry profession, where it is used to etch and clean teeth;
- In a number of consumer products, such as soaps, detergents, and toothpastes;
- As a refining and clarifying agent in the production of sugar;
- In the dyeing of cotton;

Words to Know

CATALYST A material that increases the rate of a chemical reaction without undergoing any change in its own chemical structure.

PHOSPHATE A compound that is manufactured from phosphoric acid

SUPERCOOLED Refers to a substance that remains in a liquid state at temperatures below its freezing point.

- As a binder for cement;
- In the manufacture of waxes and polishes; and
- In water and sewage treatment plants.

Phosphates (compounds made from phosphoric acid) were once used widely as "builders" in detergents. A builder is a compound that increases the cleaning efficiency of the detergent. The problem is that phosphates that escape into the natural environment can result in some undesirable changes in fresh water systems. Algae living in these systems use phosphate to grow and multiply, resulting in the conversion of fresh water lakes and ponds into swamps and bogs, and, eventually, into dry fields, a process known as eutrophication. Because of this effect, the use of phosphates in detergents has been banned in most parts of the United States.

Phosphoric acid is an extremely hazardous and toxic compound. In small amounts, it causes irritation of the skin, eyes, and respiratory system. If ingested, it can cause serious damage to the digestive system, resulting in nausea, vomiting, abdominal pain, difficulty in breathing, shock, and occasionally death by asphyxiation (suffocation). The most serious health hazards posed by phosphoric acid are of concern primarily to people who work with the product. The amount of phosphoric acid present in most domestic and household products is very small and poses little risk to users of those products.

FOR FURTHER INFORMATION

"Chemical of the Week: Phosphoric Acid." Science Is Fun. http://scifun.chem.wisc.edu/CHEMWEEK/H3PO4/H3PO4.html (accessed on October 24, 2005).

"Phosphate Primer." Florida Institute of Phosphate Research. http://www1.fipr.state.fl.us/PhosphatePrimer (accessed on October 24, 2005).

"Phosphoric Acid." DC Chemical Co., Ltd. http://www.dcchem.co.kr/english/product/p_basic/p_basic04.htm (accessed on October 24, 2005).

"Phosphoric Acid Fuel Cells." Smithsonian Institution. http://americanhistory.si.edu/fuelcells/phos/pafcmain.htm (accessed on October 24, 2005).

See Also Nitric Acid; Sulfuric Acid

Poly(Styrene-Butadiene-Styrene)

OTHER NAMES:
SBS

FORMULA:
-[CH$_2$CHC$_6$H$_5$-]$_n$-
[-CH$_2$CH=CHCH$_2$-]$_n$-
[CH$_2$CHC$_6$H$_5$-]$_n$

ELEMENTS:
Carbon, hydrogen

COMPOUND TYPE:
Organic polymer

STATE:
Solid

MOLECULAR WEIGHT:
Varies

MELTING POINT:
160°C to 200°C
(320°F to 400°F)

BOILING POINT:
Not applicable

SOLUBILITY:
Insoluble in water

KEY FACTS

OVERVIEW

Poly(styrene-butadiene-styrene) (pol-ee-STYE-reen-byoo-tah-DYE-een-STYE-reen) is a thermoplastic block copolymer of styrene and butadiene. The compound is often called simply SBS or SBS rubber. A thermoplastic polymer is one that can be converted back and forth between liquid and solid states by alternate heating and cooling. A copolymer is a polymer made from two monomers, in this case, styrene (C$_6$H$_5$CH=CH$_2$) and 1,3-butadiene (CH$_2$=CHCH=CH$_2$). The term *block copolymer* means that one section of the polymer chain consists of polystyrene (-[CH$_2$CHC$_6$H$_5$-]$_n$ to which is connected another section consisting of polybutadiene (-[-CH$_2$CH=CHCH$_2$-]$_n$), which, in turn, is connected to another section of polystyrene (-[CH$_2$CHC$_6$H$_5$-]$_n$), and so on.

A copolymer like SBS has properties of both polymers of which it is composed. In the case of SBS, the polystyrene segments give the product strength and durability, while the polybutadiene segments provide flexibility. The substance acts like natural rubber at room temperature, but becomes

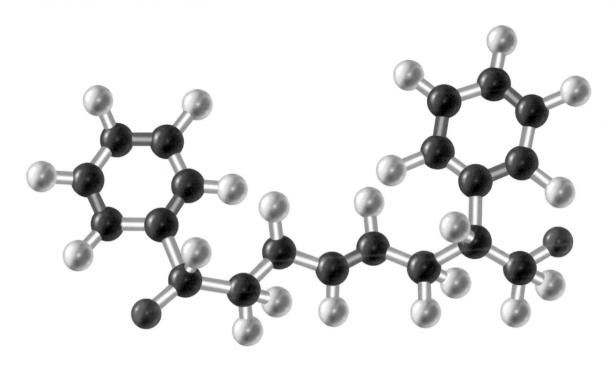

Poly-(styrene-butadiene-styrene). White atoms are hydrogen; black atoms are carbon; and turquoise atoms show where this molecule joins to other ones to form chains.
PUBLISHERS RESOURCE GROUP

soft and plastic when heated. The latter property means that products made of SBS can be formed into a variety of shapes.

SBS is resistant to abrasion and does not readily break down when exposed to heat, light, and chemicals. It may dissolve or break down when exposed to fats and oils and various types of hydrocarbon compounds and mixtures. It maintains its structure and performance well over a wide temperature range from −60°C to 150°C (−75°F to 300°F).

SBS was first developed in the early 1930s by two German chemists, Walter Bock and Eduard Tschunkur. Their research was part of Germany's Four Year Plan for self-sufficiency. Under that plan, the nation worked toward eliminating, so far as possible, the import of essential materials that might be needed in case of a war. Germans already had one type of synthetic rubber, known as *Buna*, but it had a number of disadvantages. SBS was far superior to Buna, and was soon being produced in very large amounts in German factories.

HOW IT IS MADE

Molecules of both styrene and 1,3-butadiene contain double bonds. Any compound with double bonds has the ability to form polymers. Polymerization occurs when the double bond in one monomer molecule (such as styrene) breaks apart. A hydrogen atom from a second molecule of the monomer then adds on to one end of the broken double bond. The rest of the second molecule adds to the other end of the broken double bond. A "double-molecule," consisting of two monomers joined to each other, forms. The "double-molecule" also contains a double bond. So the process can be repeated to form a "triple-molecule" consisting of three monomer molecules. The process is repeated hundreds or thousands of times producing a long chain of monomers.

In the production of a block copolymer, one extra step is added. First, a long chain of styrene monomers is produced. Then a long chain of butadiene monomers is made. Next, the two chains are joined to each other. Finally, additional chains of polystyrene and polybutadiene are added, making a very long chain consisting of alternating blocks of polystyrene and polybutadiene.

This method is used for the production of many different kinds of polymers. The most difficult problems may be (1) how to get the first few double bonds to break apart, and (2) how to stop the polymerization reaction at just the right point. The method used to make most SBS today involves the use of a butyl lithium (C_4H_9Li) catalyst, which is very effective in getting the reaction started. The reaction is terminated at some given point by the addition of dichlorodimethylsilane ($SiCl_2(CH_3)_2$). The dichlorodimethylsilane reacts with the last monomer at the end of the SBS chain, blocking the addition of any additional styrene or butadiene monomers.

COMMON USES AND POTENTIAL HAZARDS

The process by which SBS is made can be modified to produce products with somewhat different physical and chemical properties. For example, some forms of SBS are especially strong, making them suitable for the manufacture of tires, shoe soles, conveyer belts, and the tracks on caterpillar trucks. Other types of SBS are engineered to be more flexible, for use as rubber tubing, flexible toys, sporting goods, and refrigerator gaskets. SBS products can also be made in a variety of colors and shapes for use as seals, rubber mats, floor coverings, tire treads, and shoe components.

Words to Know

COPOLYMER A polymer made with two different monomers.

MONOMER A small molecular unit that joins with others to form a polymer.

POLYMER A compound consisting of very large molecules made of one or two small repeated units called monomers.

THERMOPLASTIC A material that becomes soft and moldable when heated, then hardens when it is cooled.

SBS production fluctuates or changes according to a number of factors, including market demand, the price of petroleum, and the price of natural rubber. For example, when natural rubber is readily available and inexpensive, the demand for synthetic types of rubber, such as SBS, decreases. Also, when the price of petroleum increases, SBS becomes more expensive to make and production decreases.

FOR FURTHER INFORMATION

Dick, John S., and R. A. Annicelli, eds. Rubber Technology: Compounding and Testing for Performance. Cincinnati, OH: Hanser Gardner Publications, 2001.

Johnson, Peter S. Rubber Processing: An Introduction. Cincinnati, OH: Hanser Gardner Publications, 2001.

"SBS Rubber." University of Mississippi, Polymer Science Learning Center. http://www.pslc.ws/mactest.sbs.htm (accessed on October 26, 2005).

See Also 1,3-Butadiene; Polystyrene; Styrene

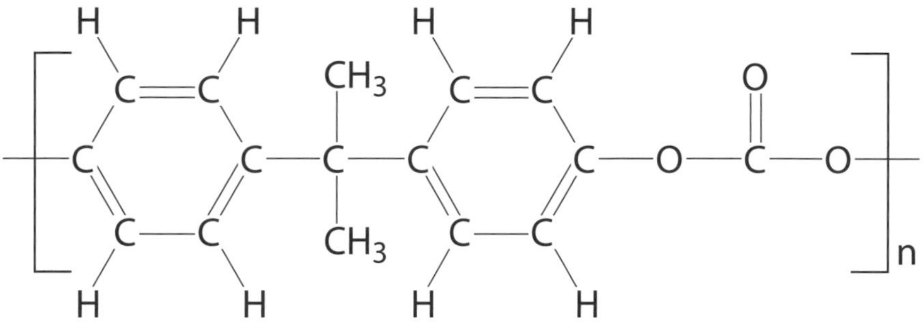

Polycarbonates

FORMULA:
Varies

ELEMENTS:
Carbon, hydrogen, oxygen

COMPOUND TYPE:
Organic polymer

STATE:
Solid

MOLECULAR WEIGHT:
Very large; varies

MELTING POINT:
Varies

BOILING POINT:
Not applicable; decomposes above melting point

SOLUBILITY:
Virtually insoluble in water

OVERVIEW

Polycarbonate (pol-ee-KAR-bun-ate) is a term used both for a specific compound and for a class of compounds with similar chemical structures. The members of this family are made by reacting phosgene ($COCl_2$) with any compound having two phenol structures. Phenol is hydroxybenzene, C_6H_5OH. The most common polycarbonate is made in the reaction between phosgene and bisphenol A ($C_6H_5OHC(CH_3)_2$-C_6H_5OH). Polycarbonates are sold under a number of trade names, including Cyrolon®, Lexan®, Markrolon®, Merlon®, Tuffak®, and Zelux®.

Polycarbonates are strong, lightweight plastics that are resistant to heat, light, chemicals, and physical shock. They are used in a number of commercial and industrial products ranging from consumer electronics to sporting goods to storage containers.

The original polycarbonate was discovered in 1953 by the German chemist Hermann Schnell (1916-1999), an employee at the Bayer AG chemical company. The product was discovered

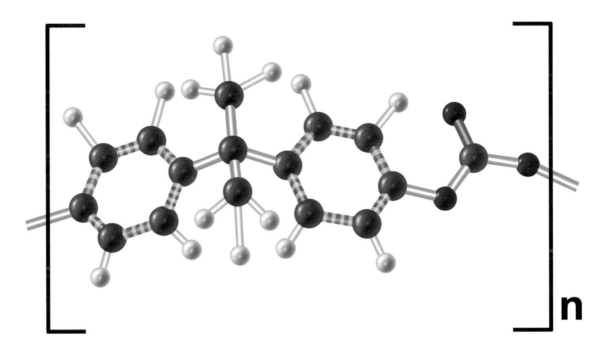

n

Polycarbonate. Red atoms are oxygen; black atoms are carbon; white atoms are hydrogen. Gray sticks indicate double bonds. Striped sticks show benzene rings. PUBLISHERS RESOURCE GROUP

almost simultaneously in the United States by chemist Daniel W. Fox (1923-1989) at the General Electric Company. Commercial use of the product began in the late 1950s, with the first large-scale production plant beginning operation in 1960. At first, polycarbonate was used primarily for electrical devices such as fuse boxes. In 1982, compact discs (CDs) were introduced and were made almost entirely of polycarbonate. Fifteen years later, polycarbonate digital videodiscs (DVDs) entered the market. In the mid-1980s, polycarbonate bottles began to replace the more cumbersome and breakable glass bottles. Today, polycarbonates are used for dozens of commercial applications.

HOW IT IS MADE

Polycarbonates are made in the reaction between phosgene and any compound containing two phenol groups, such as bisphenol A . One chlorine atom from the phosgene combines with a hydrogen atom from the hydroxyl group (-OH) in each phenol to make hydrogen chloride (HCl). As the hydrogen chloride is removed from the reaction, the phosgene remnant links with the phenol remnant. As the reaction proceeds, the phosgene-phenol grouping grows longer and

Interesting Facts

One of polycarbonate plastics most interesting uses is in the manufacture of bulletproof windows.

longer, eventually forming a large, straight-chain polymer of polycarbonate.

Other types of polycarbonates have also been made using a very different approach from that involving bisphenol A and related compounds. For example, the reaction between phosgene and allyl alcohol (CH_2=$CHCH_2OH$) produces a monomer with carbon-carbon double bonds at both ends of the molecule that can be used for polymerization. Interestingly enough, the polycarbonate produced by this process has very different physical properties from the traditional bisphenol A polymer. The allyl polymer is a clear, transparent, flexible plastic whose primary use is in the production of eyeglass lenses.

COMMON USES AND POTENTIAL HAZARDS

Polycarbonates are strong, heat resistant, and lightweight, making them ideal for applications in construction, electronics, automobiles, and appliances. Many polycarbonates are substituted for glass in safety and athletic goggles, building components, and car instrument panels because they are transparent, yet shatterproof. They are also more resistant to ultraviolet radiation than is glass. Some of the specific products made with polycarbonates include:

- Consumer electronic devices, such as cell phones, computers, pagers, and fax machines;
- Data storage devices, such as CDs and DVDs;
- Automobile parts, including tail lights, turn signals, fog lights, and headlamps;
- Appliances such as refrigerators, food mixers, hair dryers, and electric shavers;
- Safety devices, including helmets, goggles, and bulletproof windows;

Words to Know

POLYMER A compound consisting of very large molecules made of one or two small repeated units called monomers.

• Healthcare devices, such as incubators, kidney dialysis machines, and eyeglasses.

The U.S. Food and Drug Administration has accepted polycarbonates as safe for use with foods. There is little evidence that the compounds have any harmful health effects on humans. Minimal evidence exists to suggest that bisphenol A may have some harmful effects on experimental animals. But there is no evidence that similar effects occur in humans, and the amount of free bisphenol A in polycarbonates is so small as to be negligible.

FOR FURTHER INFORMATION

Lazear, N. R. "Polycarbonate: High-Performance Resin." Advanced Materials & Processes (February 1995): 43-45.

"Polycarbonate." Association of Plastics Manufacturers in Europe. http://www.plasticseurope.org/content/default.asp?PageID=43 (accessed on October 24, 2005).

"Polycarbonate (Makrolon®, Lexan®, Zelux®)." San Diego Plastics. http://www.sdplastics.com/polycarb.html (accessed on October 24, 2005).

"Polycarbonate Plastics." Bisphenol A. http://www.bisphenol-a.org/human/polyplastics.html (accessed on October 24, 2005).

"Polycarbonates." Polymer Science Learning Center. http://www.pslc.ws/mactest/pc.htm (accessed on October 24, 2005).

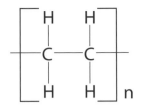

OTHER NAMES:
Ethene homopolymer

FORMULA:
$-[-CH_2-CH_2-]-_n$

ELEMENTS:
Carbon, hydrogen

COMPOUND TYPE:
Organic polymer

STATE:
Solid

MOLECULAR WEIGHT:
1,500 to 100,000 g/mol

MELTING POINT:
Varies: about 85°C -110°C (185°F-230°F)

BOILING POINT:
Not applicable; decomposes above melting point

SOLUBILITY:
Insoluble in water and most organic solvents; soluble in hydrocarbons and halogenated hydrocarbons

Polyethylene

OVERVIEW

Polyethylene (pol·ee·ETH·uh·leen) is a thermosetting white solid resistant to high temperatures, most inorganic and organic chemicals, and physical impact. It is also an electrical non-conductor. A thermosetting polymer is one that, once it is melted and formed, can not be re-melted. Polyethylene is available in a variety of forms, the most common of which are high-density (HD or HDPE), low density (LD or LDPE), linear low density (LLD or LLDPE) and cross-linked (CLPE). These forms of the compound differ with respect to the structure of the poly-ethylene chains and their relationship to each other. For example, if all of the polyethylene chains are straight chains without branches, they can pack together tightly forming a high density product. By contrast, low density polyethylene consists of shorter chains with many side branches on them. The side branches prevent adjacent polymer chains from getting too close to each other. In cross-linked polyethylene, adjacent polymer chains actually form chemical bonds with each other, holding them in a regular, almost crystalline pattern.

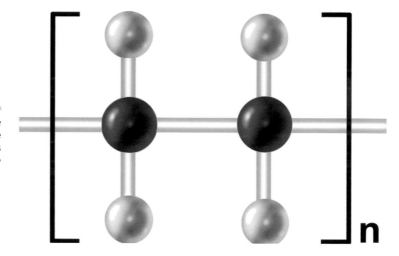

Polyethylene. White atoms are hydrogen and black atoms are carbon. PUBLISHERS RESOURCE GROUP

Polyethylene was first prepared accidentally in 1889 by the German chemist Hans von Pechmann (1850-1904). Von Pechmann was heating diazomethane ($H_2C=N=N$) when he observed the formation of a white, waxy solid. When his colleagues identified the substance as containing repeated methylene ($-CH_2$) units, they called the material polymethylene. Von Pechmann did not pursue his discovery, nor did any of his colleagues until the 1930s. Then, in 1933, two chemists at the English firm of Imperial Chemicals Industries (ICI), Reginald Gibson and Eric Fawcett, accidentally re-discovered polyethylene. While attempting to pressurize a mixture of ethylene and benzaldehyde, Gibson and Fawcett observed the formation of a white waxy solid within the pressure vessel. At first, they were unable to account for this reaction. Eventually, however, they discovered that a tiny hole in the pressure vessel had allowed oxygen to seep into the tank, catalyzing the conversion of ethylene to polyethylene. Two years later, two other ICI chemists, J. C. Swallow and M. W. Perrin duplicated the Gibson-Fawcett experiment and devised an efficient commercial method for making the polymer.

Beginning in the 1950s, researchers in a number of countries began to explore the use of catalysts to increase the efficiency of reactions by which polyethylene is made. For example, Robert Banks and John Hogan at Phillips Petroleum invented a procedure using chromium trioxide (Cr_2O_3) for the preparation of high-density polyethylene and another form of the product known as crystalline polyethylene. In 1953, the German chemist Karl Ziegler (1898-1973)

Interesting Facts

Phillips Petroleum had difficulty maintaining quality control in the early years of making polyethylene. As a result, it produced large quantities of the product that could not be sold for commercial, industrial, or household use. The company faced financial ruin. Fortunately, a new toy came into existence in the mid-1950s, the hula-hoop. A hula-hoop is simply a ring of plastic that could be made with low-grade polyethylene. The first company to market hula-hoops, called Wham-O, bought up much of Phillips' defective polyethylene stock for its hula-hoops. The company sold more than twenty million hula-hoops in its first six months of existence at a cost of $1.98 each. Wham-O was a booming success, and it saved Phillips from financial ruin.

used titanium halides and organoaluminum compounds to make polyethylene under even lower temperatures and pressures. And in 1976, the German chemists Walter Kaminsky (1941-) and Hansjörg Sinn (1929-) invented a third method of production using metallocenes (organic compounds that contain a metal) as a catalyst.

HOW IT IS MADE

Polyethylene is made by polymerizing ethylene (ethene; $CH_2=CH_2$). Polymerization occurs when the double bond in ethylene breaks, allowing one molecule of ethylene to combine with a second molecule of ethylene: $CH_2=CH_2 + CH_2=CH_2 \rightarrow CH_3CH_2CH=CH_2$. The product of that reaction also contains a double bond, allowing the reaction to be repeated: $CH_3CH_2CH=CH_2 + CH_2=CH_2 \rightarrow CH_3CH_2CH_2CH_2CH=CH_2$. Once again, the final product contains a double bond, and the reaction can be repeated again and again and again.

Polymerization occurs when some outside agent provides the energy to break the double bond in ethylene to get the reaction started. Heat, light, ultraviolet radiation, a beam of electrons, and gamma rays have all been used to initiate polymerization. Polymerization occurs at lower energy levels, and it has been the search for catalysts to achieve this objective

Words to Know

CATALYST A material that increases the rate of a chemical reaction without undergoing any change in its own chemical structure.

POLYMER A compound consisting of very large molecules made of one or two small repeated units called monomers.

that have led to the processes developed by Banks and Hogan, Ziegler, Kaminsky and Sinn, and other researchers.

COMMON USES AND POTENTIAL HAZARDS

An estimated 55 million metric tons (60 million short tons) of polyethylene are produced worldwide each year. In the United States, an estimated 7 million metric tons (8 million short tons) of high-density polyethylene will be produced in 2006 and an estimated 2.7 million metric tons (3.0 million short tons) of low-density polyethylene will be made. The greatest fraction of HDPE is used in the manufacture of molded products, film and sheeting, pipes and tubing, fibers, and gasoline and oil containers. The most important application of LDPE is in the manufacture of packaging films for foods as well as coatings that are sprayed or otherwise applied to all kinds of surfaces. It is also used widely to make liners for drums and other shipping containers, wire and cable coating, trash bags, squeeze bottles, inexpensive dinnerware, drop cloths, swimming pool covers, toys, and electric insulation.

No health hazards from polyethylene in any form have yet been identified.

FOR FURTHER INFORMATION

Meikle, Jeffrey L. *American Plastic: A Cultural History.* http://www.ilo.org/public/english/protection/safework/cis/products/icsc/dtasht/_icsc14/icsc1488.htm (accessed on October 24, 2005).

"Polyethylene." In *World of Invention.* 2nd ed. Edited by Kimberley A. McGrath and Bridget Travers. Detroit, MI: Gale, 1999.

"Polyethylene Specifications." Boedeker Plastics. http://www.boedeker.com/polye_p.htm (accessed on October 24, 2005).

See Also Polypropylene

$$\left[-CH_2 - \underset{\underset{\underset{H}{|}}{\underset{O}{|}}{\overset{\overset{CH_3}{|}}{C}} - CH_2 - \right]_n$$

OTHER NAMES:
Acrylic, PMMA

FORMULA:
-[-CH$_2$C(CH$_3$)
(COOH)CH$_2$-]-$_n$

ELEMENTS:
Carbon, hydrogen,
oxygen

COMPOUND TYPE:
Organic polymer

STATE:
Solid

MOLECULAR WEIGHT:
Varies: 250,000 to
over 1,000,000 g/mol

MELTING POINT:
Varies: usually above
100°C (200°F)

BOILING POINT:
Not applicable

SOLUBILITY:
Insoluble in water;
best solvents are
mixtures of two or
more organic
solvents, aromatic
hydrocarbons,
halogenated hydro-
carbons, and
tetrahydrofuran

KEY FACTS

Polymethyl Methacrylate

OVERVIEW

Polymethylmethacrylate (POL-ee-meth-uhl-meth-AK-rill-ate) is a clear thermoplastic resin used to make windshields, visors, coatings for baths, advertising signs, and contact lenses. It is also widely used in dentistry and medicine. A thermoplastic resin is one that becomes soft when heated and hard when cooled. It can be converted back and forth any number of times between the solid and liquid states by further heating and cooling.

Polymethylmethacrylate (PMMA) is more transparent than glass and six to seventeen times more resistant to breakage than glass. Another advantage it has over glass is that, if it does break, it falls apart into dull-edged pieces. PMMA is resistant to water, inorganic acids and bases, but is vulnerable to many organic solvents.

PMMA was first synthesized in 1928 in the laboratories of the German chemical firm Röhm and Haas. After five years of research, one of the firm's founders, Otto Röhm, found a way of manufacturing sheets of polymethylmethacrylate. He patented his invention and the company was soon producing the

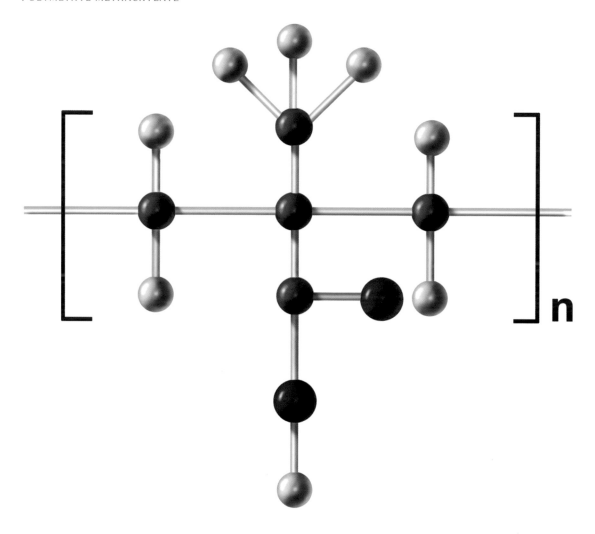

Poly(methyl methacrylate).
Red atoms are oxygen; white
atoms are hydrogen; black
atoms are carbon. Gray stick
shows a double bond.
PUBLISHERS RESOURCE GROUP

product under the trade name of Plexiglas®. At about the same time, the product was being developed independently in the United States by the Dow chemical company, who sold its product under the trade name of Lucite®. Over time, a number of other chemical companies produced their own version of PMMA under a variety of trade names, including Acrylite®, Acrypet®, Crinothene®, Degalan®, Diakon®, Elvacite®, Kallocryl®, Metaplex®, Osteobond®, Paraglas®, Perspex®, Pontalite®, Sumipex®, Superacryl®, and Vedril®.

HOW IT IS MADE

The monomer from which polymethylmethacrylate is made is methyl methacrylate, $CH_2=C(CH_3)COOCH_3$. Methyl

Interesting Facts

- Polymethylmethacrylate exhibits a phenomenon known as total internal reflection. That term means that a light beam transmitted through a solid tube made of PMMA reflects off the inner surface of the tube. This property allows a light beam to be transmitted around corners and bends and out the end of a tube made of PMMA.

- A black-light reactive tatoo ink made from PMMA and microspheres of fluorescent dye is used on wildlife to track activities such as migration and growth patterns.

- The spectator shield in hockey arenas is made of PMMA.

- Plexiglas® was exhibited by Röhm & Haas at the World's Trade Fair in Paris in 1937. One of the exhibits was a transparent violin made from the product.

methacrylate is made in the reaction between acetone cyanohydrin ($(CH_3)_2COHCN$) and methanol (methyl alcohol; CH_3OH) in the presence of a sulfuric acid catalyst. As with all polymers, methyl methacrylate can be polymerized by a number of agents, including heat, radiation, and certain chemicals known as free-radical initiators. Once polymerization of methyl methacrylate begins, it can be processed in a number of ways. One of the most common processing system used involves passing the liquid material between two polished stainless steel belts. The distance between the belts is set to the desired thickness for the acrylic sheet. The acrylic is cured by a process of cooling and heating after leaving the steel belt. The final product is then cut to desired lengths at the end of the production line.

COMMON USES AND POTENTIAL HAZARDS

The most common use for PMMA is as a glass substitute. PMMA offers many benefits over glass because it is more transparent, less dense, stronger, and shatterproof. In ophthalmology (the branch of medicine that deals with diseases of the eye and their treatment), PMMA is used to make replacement lenses for the eye when the original lens has been removed for

Words to Know

CATALYST a material that increases the rate of a chemical reaction without undergoing any change in its own chemical structure.

MONOMER one of the small, relatively simple molecules from which polymers are made.

some reason, such as the growth of a cataract. It is also used to make hard contact lenses. Other medical and dental applications of the material include its use as a dental cement, for the manufacture of bases and linings for dentures (false teeth), and to make a bone cement used in the reconstruction of broken or damaged bones and to fix implants in place.

Powdered polymethylmethacrylate presents health hazards because it may cause irritation of the skin, eyes, and respiratory tract. This hazard is of concern to people who work with the material in its raw form.

The glass-like material with which most people come into contact poses no health hazard for humans.

FOR FURTHER INFORMATION

"Material Information Polymethylmethacrylate, PMMA, Acrylic." Goodfellow. http://www.goodfellow.com/csp/active/static/E/ME30.HTML (accessed on October 24, 2005).

Meikle, J. L. *American Plastic: A Cultural History.* New Brunswick, N.J.: Rutgers University Press, 1997.

"More on the Manufacturing of Acrylic Sheet." *Plastics Distributor & Fabricator Magazine* (November/December 2000). Available online at http://www.plasticsmag.com/features.asp?fIssue=Nov/Dec-00&aid=3053 (accessed on October 24, 2005).

"Polymethylmethacrylate." Kids' Macrogalleria. University of Southern Mississippi, The Polymer Science Learning Center. http://www.pslc.ws/macrog/kidsmac/pmma.htm (accessed on October 24, 2005).

"Plexiglas® Primer." Ridout Plastics. http://www.ridoutplastics.com/plexprim.html (accessed on October 24, 2005).

$$\left[\begin{array}{cc} \overset{\displaystyle H}{\underset{\displaystyle H}{\vert}} C & \overset{\displaystyle H}{\underset{\displaystyle CH_3}{\vert}} C \\ \end{array} \right]_n$$

OTHER NAMES:
Propylene polymer;
I-propene
homopolymer

FORMULA:
$-[-CH(CH_3)CH_2-]-_n$

ELEMENTS:
Carbon, hydrogen

COMPOUND TYPE:
Organic polymer

STATE:
Solid

MOLECULAR WEIGHT:
Very large; 40,000 g/
mol and up

MELTING POINT:
Varies: 165°C–170°C
(330°F 340°F)

BOILING POINT:
Not applicable

SOLUBILITY:
Insoluble in water and
cold organic solvents;
softens in warm
organic solvents, but
does not dissolve;
soluble in hydrocar-
bons and halogenated
hydrocarbons

KEY FACTS

Polypropylene

OVERVIEW

Polypropylene (pol-ee-PRO pih-leen) is a translucent white solid that is resistant to attack by heat, abrasion, inorganic acids and bases, most organic materials, and bacteria and fungi. It has high electrical resistance and tensile strength (its resistance to being pulled apart) and is very flexible. It takes on color well and can be coated with chrome. It can be prepared in a number of shapes and forms by extrusion and molding. Polypropylene is a thermoplastic polymer, meaning that it can be heated and cooled repeatedly, changing from a solid to a liquid and back again. Polypropylene is currently available under a number of trade names, including Amco®, Amerfil®, Azdel®, Beamette®, Clysar®, Daplen®, Dexon®, Epolene®, Gerfil®, Herculon®, Lambeth®, Lupareen®, Meraklon®, Mitsui Polypro®, Noblen®, Novolen®, Pellon®, Polypro®, Profax®, Propathene®, Propolin®, Propophane®, Shoallomer®, and Tuff-Lite®. Considerable dispute exists as to who should receive credit for inventing propylene. According to one history of the compound, it

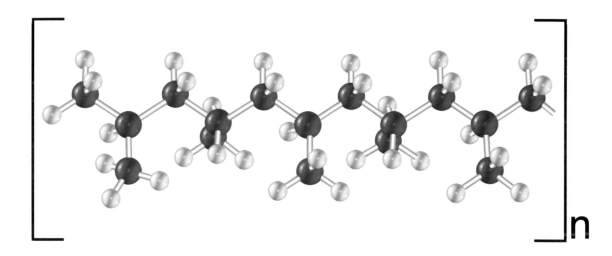

Polypropylene. White atoms are hydrogen and black atoms are carbon. PUBLISHERS RESOURCE GROUP

was discovered independently about nine times. Patent disputes over the discovery lasted from the 1950s to 1989, when official credit was finally given to two researchers at the Phillips Petroleum Company, J. Paul Hogan (1919-) and Robert Banks (1921-1989). Phillips began selling polypropylene in 1951 under the trade name of Marlex®.

Much of the early research on polypropylene was conducted by the Italian chemist Giulio Natta (1903-1979), then an employee at the Italian chemical firm of Montecatini. While working earlier on the development of a related compound, polyethylene, Natta discovered some of the fundamental principles that governed the successful commercial production of polymers. In 1957, Montecatini began producing its own version of polypropylene. Six years later, Natta and his colleague, German chemist Karl Ziegler (1898-1973) shared the Nobel Prize in chemistry for their research on polymers.

HOW IT IS MADE

Polypropylene is made by the polymerization of propylene (propene; $CH_3CH=CH_2$). Polymerization is the process by which a single monomer unit (propylene in this case) is added to a second monomer of the same kind. The procedure is then repeated over and over again. Each time another monomer is added to the growing chain, the molecule gets larger and larger. Normally, polymerization is initiated by any of a number of agents, including radiation, light, or heat.

Interesting Facts

- Natta applied for a patent for polypropylene before telling Ziegler of his success in making the compound. Ziegler was so angry that the two men did not speak for many years. They reconciled only when they were both awarded the Nobel Prize at the same time.

- Australian bank notes (paper money) are made from polypropylene, which makes them more durable than paper bills.

Polymerization of propylene presents a somewhat different problem, however, because of the presence of methyl ($-CH_3$) groups extending off the main chain of the molecule ($-[-CH(CH_3)CH_2-]-_n$). If polymerization is allowed to proceed on its own, some methyl groups will extend in one direction from the main chain, and others in a different direction. The product of this reaction is an amorphous product, one without crystalline shape, that has only a few very limited uses. To produce crystalline polypropylene, with all the desirable properties noted above, polymerization must be controlled to make sure that all methyl groups are on the same side of the main chain. One of Natta and Ziegler's great contributions was the discovery of catalysts capable of achieving the correct orientation of methyl groups. They found that metal halides, such as titanium chloride, could produce this effect. More recently, the German chemist Walter Kaminsky (1941–) and his colleagues found another group of catalysts that could polymerize crystalline polypropylene even more efficiently, a group of compounds called metallocenes. The manufacture of polypropylene today depends heavily on the use of such catalysts.

COMMON USES AND POTENTIAL HAZARDS

About one-third of all the polypropylene consumed in the United States is used to make fibers, for the manufacture of products such as blankets, fabrics, carpets, yarns, fish nets, protective clothing, laundry bags, and ropes. The next largest

Words to Know

CATALYST a material that increases the rate of a chemical reaction without undergoing any change in its own chemical structure.

POLYMER a compound consisting of very large molecules made of one or two small repeated units called monomers.

uses are in the production of rigid packaging materials, such as crates, food containers, and bottles; in household products, such as dishes, bowls, outdoor carpeting, and outdoor furniture; and in packaging film. Hospitals use many surgical objects made out of polypropylene, taking advantage of its low cost and ability to be sterilized. Automobile manufacturers use the compound almost everywhere on the body of their cars. About 20 percent of all the polypropylene produced is used to make a large variety of products, including wire and cable insulation, medical tubing, pipe fittings, battery cases, drinking straws, and packaging foam.

No human health hazards have been identified for polypropylene in the form in which most people come into contact with the compound.

FOR FURTHER INFORMATION

Karger-Kocsis, J. *Polypropylene—An A-Z Reference.* New York: Springer-Verlag, 1998.

Meikle, Jeffrey L. *American Plastic: A Cultural History.* Piscataway, N.J.: Rutgers University Press, 1997.

"Polypropylene." The Macrogalleria. Polymer Learning Center, University of Southern Mississippi. http://www.pslc.ws/macrogcss/pp.html (accessed on October 24, 2005).

"Polypropylene." *World of Chemistry.* Edited by Robyn V. Young. Detroit, Mich.: Gale, 1999.

See Also Polyethylene

Polysiloxane

OVERVIEW

The term polysiloxane (pol-ee-sill-OK-sane) refers to a class of compounds whose molecules consist of a silicon-oxygen backbone (-Si-O-Si-O-Si-O-Si)$_n$ arranged either in a linear or cyclic (ring) pattern. Each silicon in the chain has two additional oxygen atoms attached to it. In many cases, the polysiloxanes also have one or more alkyl groups attached to the main chain replacing one or more of the oxygens. An alkyl group is an alkane, a saturated hydrocarbon, lacking one hydrogen atom. Examples of alkyl groups are the methyl (-CH_3) and ethyl (-CH_2CH_3) groups. In one common type of siloxane, all of the oxygens that are not a part of the backbone of the chain are replaced by methyl groups. Polysiloxanes that contain alkyl groups are known as organosiloxanes or, more commonly, silicones.

Researchers have now learned how to modify organosiloxane polymers by using various alkyl groups in chains of various lengths and conformations to produce a very wide array of products. All are organosiloxanes, but with very

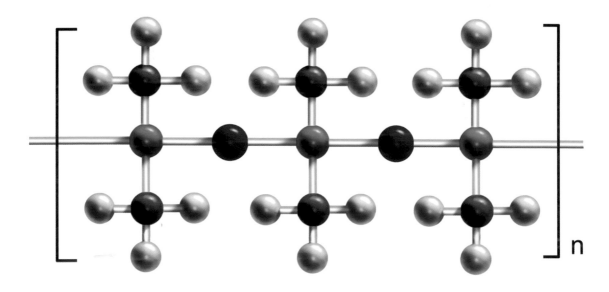

n

Polysiloxane. Red atoms are oxygen; white atoms are hydrogen; black atoms are carbon; and turquoise atoms are silicon. PUBLISHERS RESOURCE GROUP

different physical and chemical properties and, hence, with very different applications. They range from liquid to gel to a semi-solid, rubber-like material. In general, the orgaosiloxanes tend to be chemically inert and resistant to heat. Although not identical, the terms polysiloxane, siloxane, and organosiloxane, and silicone are often used interchangeably.

The chemistry of silicon has long fascinated chemists. The element lies just below carbon in the periodic table, meaning that its physical and chemical properties are similar to carbon from which millions of organic compounds are made. The hope has long been to discover if silicon also can make a diverse number of compounds, as carbon can. One researcher who pursued this question in the early twentieth century was the English chemist Frederic Stanley Kipping (1863-1949). Kipping made use of a new kind of chemical reaction, called the Grignard reaction, after the French chemist Victor Grignard (1871-1935) who invented it. With the Grignard reaction, Kipping was able to make a number of silicon compounds to which were bonded alkyl groups in a variety of conformations. These compounds were the first organosiloxanes produced and studied in any detail, earning Kipping the title of "Father of Silicone Chemistry." During his lifetime, Kipping wrote more than 50 scholarly papers on his work. He did not, however, see the possibility of commercial applications for his discoveries.

The commercialization of silicones did not become possible until the early 1940s when a research chemist at the General Electric Company, E. G. Rochow (1909–2002) found an efficient way of making organosiloxanes in large quantities. Rochow's discovery came at an opportune time, the beginning of World War II. A number of military applications were found for the new product almost immediately. Within five years, a number of major chemical companies, including General Electric, Dow-Corning, Union Carbide, Stauffer Chemical, Wacher-Chemie, and Farbenfabriken Bayer A. G., had begun making organosiloxanes in large quantities.

HOW IT IS MADE

A variety of production methods is available for making different types of organosiloxanes. Probably the most popular approach begins with the reduction of silicon dioxide (sand; SiO_2) in an electric furnace. In that reaction, the silicon dioxide breaks down into silicon and oxygen: $SiO_2 \rightarrow Si + O_2$. The silicon is then treated with methyl chloride (CH_3Cl), which produces a mixture of methylchlorosilanes. Methylchlorosilanes are compounds that contain a single silicon atom to which are attached one or more methyl groups and one or more chlorine atoms. The methylchlorosilane produced in largest amount is dimethyldichlorosilane, $(CH_3)_2SiCl_2$. When this compound is treated with water, all of the chlorines are replaced by oxygens and the resulting compound polymerizes spontaneously. That is, individual molecules of the newly-formed silicon-carbon-oxygen compound begin to react with each other to form long organosiloxane chains. The size and character of the chain can be controlled by adding compounds that stop the polymerization reaction at some given point, resulting in the formation of a liquid, a gel, a semi-solid, or some other form of the product.

COMMON USES AND POTENTIAL HAZARDS

Organosiloxanes are very heat resistant, so they do not easily melt, like most other organic compounds. They are also water-repellant and can withstand extremes of sunlight, moisture, cold, and attack by most chemicals. These properties make them useful for protective coatings, electrical insulation, adhesives, lubricants, paints, and rubber-like materials. Some silicones are also used to make nonstick

Interesting Facts

- One of the most famous footprints in the world—that made by Neil Armstrong during his landing on the Moon in 1969—was made with a boot with a silicone rubber sole.

- Silicone was first used for breast implants in the 1960s for women who had undergone mastectomies, surgical removal of their breasts. Silicone implants later became popular with women who had no medical problems, but wanted larger breasts. By the 1980s, many women with breast implants began complaining of pain, chronic fatigue, inflammation of breast tissue, and other medical problems thought to be associated with their silicone breast implants. Some of these women sued Bristol-Meyers Squibb, Dow Chemical, 3M, and other companies who made silicone for breast implants. By 1995, more than 20,000 legal claims had been brought against Dow alone. Faced with this staggering number of claims, the company went bankrupt.

surfaces, such as pans and spatulas for the kitchen. They are also used in making other kitchen items, such as oven mitts, because of their heat resistance.

Methyl silicones are also a major ingredient in personal care products. They are added to shaving lotions to provide lubrication and to give these products a non-greasy, yet silky, feeling. They help hair-styling products to spread more easily, and they increase the skin protection factor (SPF) in sunscreens. Silicones are also used in deodorants, perfumes, and nail polishes.

Some industrial and commercial applications of silicones include:

- As adhesives and sealants in many aircraft parts, including doors, windows, wings, fuel tanks, hydraulic switches, overhead bins, wing edges, leading gear electrical devices, vent ducts, engine gaskets, electrical wires and black boxes;

Words to Know

ALKANE a hydrocarbon where all the bonds between atoms are single bonds, the carbons are linked in a chain (except for methane, CH_4), and every carbon atom is saturated, or bonded to the maximum number of hydrogen atoms.

ALKYL GROUP an alkane that is lacking one hydrogen atom.

HYDROCARBON a chemical that consists of carbon and hydrogen.

- In the construction business as sealants for all kinds of building materials, including concrete, glass, granite, steel and plastic;

- In the manufacture of many automobile parts, including air bags, gaskets, headlamps, hydraulic bearings, ignition cables, radiator seals and hoses, shock absorbers, spark plug boots and ventilation flaps;

- As heat transfer agents;

- For the weatherproofing of concrete and other surfaces;

- In the production of surgical membranes and implants;

- In a great variety of electrical and electronic appliances, devices, and parts, including computer cases, keyboards, copy-machine components, and telephones; and

- For providing attractive and sturdy finishes for fabric and clothing.

FOR FURTHER INFORMATION

"The Basics of Silicon Chemistry." Dow Chemical Company. http://www.dowcorning.com/content/sitech/sitechbasics/silicones.asp (accessed on October 26, 2005).

"Heat and Chemical Resistant Silicone Rubber. Silicones 2. Organic Silicon Chemistry." ChemCases.com Kennesaw State University. http://www.chemcases.com/silicon/sil2cone.htm (accessed on October 26, 2005).

"Silicones." Macrogalleria, University of Southern Mississippi. http://www.pslc.ws/macrogcss/silicone.html (accessed on October 26, 2005).

"Silicones Science On-Line." Centre Européen des Silicones. http://www.silicones-science.com/ (accessed on October 26, 2005).

"What Are Silicones?" Silicones Environmental, Health and Safety Council of North America. http://www.sehsc.com/index.asp (accessed on October 26, 2005).

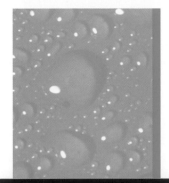

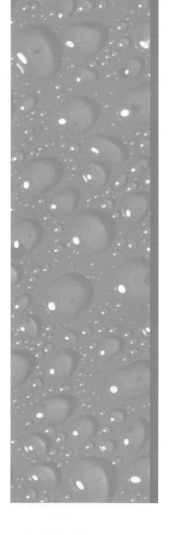

appendices

Lists of compounds

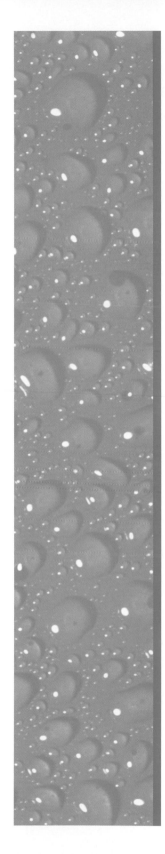

Compounds by Formula

AgI
Silver Iodide

AgNO$_3$
Silver Nitrate

Ag$_2$O
Silver(I) Oxide

Ag$_2$S
Silver(I) Sulfide

AlF$_3$
Aluminum Fluoride

Al(OH)$_3$
Aluminum Hydroxide

Al$_2$O$_3$
Aluminum Oxide

CCl$_2$F$_2$
Dichlorodifluoromethane

CCl$_4$
Carbon Tetrachloride

-[-CF$_2$-]-$_n$
Polytetrafluoroethylene

CH≡CH
Acetylene

-[-CH(CH$_3$)CH$_2$-]-$_n$
Polypropylene

CHCl$_3$
Chloroform

CH$_2$O
Formaldehyde

-[-CH$_2$C(CH$_3$)(COOH)CH$_2$-]-$_n$
Polymethyl Methacrylate

CH$_2$=C(CN)COOCH$_3$
Cyanoacrylate

CH$_2$=CHCH=CH$_2$
1,3-Butadiene

CH$_2$=CHCH$_3$
Propylene

CH$_2$=CH(CH$_3$)CH=CH$_2$
Isoprene

-[-CH$_2$ CH(COONa)-]$_n$-
Sodium Polyacrylate

-[CH$_2$CHC$_6$H$_5$-]$_n$-
[-CH$_2$CH=CHCH$_2$-]$_n$-[CH$_2$CHC$_6$H
Poly(Styrene-Butadiene-
Styrene)

-[-CH$_2$CHCl-]-$_n$
Polyvinyl Chloride

CH$_2$=CH$_2$
Ethylene

(CH$_2$CH$_2$Cl)$_2$S
2,2'-Dichlorodiethyl Sulfide

-[-CH_2-CH_2-]-$_n$
Polyethylene

-[-$CH_2C_6H_5$-]-$_n$
Polystyrene

$CH_2NO_2CHNOCH_2NO_2$
Nitroglycerin

$CH_2OHCHOHCH_2OH$
Glycerol

CH_2OHCH_2OH
Ethylene Glycol

$(CH_2)_2O$
Ethylene Oxide

$CH_3CHOHCH_3$
Isopropyl Alcohol

$CH_3CHOHCOOH$
Lactic Acid

CH_3CH_2OH
Ethyl Alcohol

CH_3COCH_3
Dimethyl Ketone

$CH_3CONHC_6H_4OH$
Acetaminophen

$CH_3COOCH_2CH_2CH(CH_3)_2$
Isoamyl Acetate

$CH_3COOC_2H_5$
Ethyl Acetate

$CH_3COOC_5H_{11}$
Amyl Acetate

$CH_3COOC_6H_4COOH$
Acetylsalicylic acid

CH_3COOH
Acetic acid

$CH_3C_5HN(OH)(CH_2OH)_2$
Pyridoxine

$CH_3C_6H_9$ $(C_3H_7)OH$
Menthol

CH_3OH
Methyl Alcohol

$(CH_3O)(OH)C_6H_3CHO$
Vanillin

CH_3SH
Methyl Mercaptan

$(CH_3)_2CHCH_2CH_2NO_2$
Amyl Nitrite

$C_5H_{11}NO_2$
Amyl Nitrite

$(CH_3)_2C_5H_3NSO(COOH)NHCOR$
Penicillin

$(CH_3)_3COCH_3$
Methyl-t-butyl Ether

CH_4
Methane

CH_4S
Methyl Mercaptan

CO
Carbon Monoxide

-[-$CO(CH_2)_4CO$-$NH(CH_2)_6NH$-]-$_n$
Nylon 6 and Nylon 66

-[-$CO(CH_2)_5NH$-]-$_n$
Nylon 6 and Nylon 66

-[-$CONH$-C_6H_4-$NCOO$-CH_2CH_2-O-]-$_n$
Polyurethane

$COOH(CH_2)_2CH(NH_2)COONa$
Monosodium Glutamate

CO_2
Carbon Dioxide

C_2H_2
Acetylene

$C_2H_2O_4$
Oxalic Acid

$[C_2H_3Cl]_n$
Polyvinyl Chloride

C_2H_4
Ethylene

$[C_2H_4]_n$
Polyethylene

C_2H_4O
Ethylene Oxide

$C_2H_4O_2$
Acetic acid

C_2H_6O
Ethyl Alcohol

$C_2H_6O_2$
Ethylene Glycol

$[C_3H_3O_2Na]_n$
Sodium Polyacrylate

$C_3H_5N_3O_5$
Nitroglycerin

$[C_3H_6]_n$
Polypropylene

C_3H_6O
Dimethyl Ketone

$C_3H_6O_3$
Lactic Acid

C_3H_8
Propane

C_3H_8O
Isopropyl Alcohol

$C_3H_8O_3$
Glycerol

C_4H_6
1,3-Butadiene

$C_4H_8Cl_2S$
2,2'-Dichlorodiethyl Sulfide

$C_4H_8O_2$
Ethyl Acetate

C_4H_{10}
Butane

$C_4H_{10}S$
Butyl Mercaptan

$C_5H_4NC_4H_7NCH_3$
Nicotine

$C_5H_5NO_2$
Cyanoacrylate

C_5H_8
Isoprene

$[C_5H_8]_n$
Polymethyl Methacrylate

$C_5H_8NNaO_4$
Monosodium Glutamate

$C_5H_{12}O$
Methyl-t-butyl Ether

$C_6H_2(CH_3)(NO_2)_3$
2,4,6-Trinitrotoluene

$C_6H_3ClOH\cdot O\cdot C_6H_3Cl_2$
Triclosan

$C_6H_3Cl_2NHCONHC_6H_4Cl$
Triclocarban

$C_6H_5CH=CHCHO$
Cinnamaldehyde

$C_6H_5CH-CH_2$
Styrene

$C_6H_5CH(CH_3)_2$
Cumene

$C_6H_5CH_3$
Toluene

C_6H_5COOH
Benzoic Acid

$C_6H_5C_2H_5$
Ethylbenzene

$C_6H_5NO_2$
Niacin

C_6H_5OH
Phenol

C_6H_6
Benzene

$C_6H_6Cl_6$
Gamma-1,2,3,4,5,6-Hexa-
chlorocyclohexane

C_6H_6O
Phenol

$[C_6H_7O_2(OH)_2OCS_2Na]_n$
Cellulose Xanthate

$C_6H_8O_6$
Ascorbic Acid

$C_6H_8O_7$
Citric Acid

$(C_6H_{10}O_5)_n$
Cellulose

$[C_6H_{11}NO]_n$
Nylon 6 and Nylon 66

$C_6H_{11}NHSO_3Na$
Sodium Cyclamate

$C_6H_{12}NNaSO_3$
Sodium Cyclamate

$C_6H_{12}O_2$
Butyl Acetate

$C_6H_{12}O_6$
Fructose
Glucose

C_6H_{14}
Hexane

$C_6H_4(CH_3)CON(C_2H_5)_2$
N,N-Diethyl-3-Methylbenza-
mide

$C_7H_5NO_3S$
Saccharin

$C_7H_5N_3O_6$
2,4,6-Trinitrotoluene

$C_7H_6O_2$
Benzoic Acid

$[C_7H_7]_n$
Polystyrene

$C_8H_7N_3O_2$
Luminol

$C_7H_8N_4O_2$
Theobromine

$[C_7H_{11}NaO_5S_2]_n$
Chloroform

$C_7H_{14}O_2$
Isoamyl Acetate
Amyl Acetate

C_8H_8
Styrene

$[C_8H_8]_n\cdot[C_4H_6]_n\cdot[C_8H_8]_n$
Poly(Styrene-Butadiene-
Styrene)

$C_8H_9NO_2$
Acetaminophen

C_8H_{10}
Ethylbenzene

$C_8H_{10}N_4O_2$
Caffeine

$C_8H_{11}NO_3$
Pyridoxine

C_9H_8O
Cinnamaldehyde

$C_9H_8O_4$
Acetylsalicylic acid

C_9H_{12}
Cumene

$C_{10}H_8$
Naphthalene

$C_{10}H_{14}N_2$
Nicotine

$C_{10}H_{16}O$
Camphor

$C_{10}H_{20}O$
Menthol

$C_{11}H_{16}O_2$
Butylated Hydroxyanisole
and Butylated Hydroxyto-
luene (BHA)

$C_{12}H_7Cl_3O_2$
Triclosan

$C_{12}H_{16}N_4O_{18}$
Cellulose Nitrate

$C_{12}H_{17}ClN_4OS$
Thiamine

$C_{12}H_{17}NO$
N,N-Diethyl-3-Methylbenza-
mide

$[C_{12}H_{22}N_2O_2]_n$
Nylon 6 and Nylon 66

$C_{12}H_{22}O_{11}$
Lactose
Sucrose

$C_{13}H_9Cl_3N_2O$
Triclocarban

$C_{13}H_{18}O_2$
2-(4-Isobutylphenyl)propionic
Acid

$C_{14}H_9Cl_5$
Dichlorodiphenyltrichloro-
ethane

$C_{14}H_{14}O_3$
Naproxen

$C_{14}H_{18}N_2O_5$
L-Aspartyl-L-Phenylalanine
Methyl Ester

$C_{15}H_{24}O$
Butylated Hydroxyanisole
and Butylated Hydroxyto-
luene (BHT)

$C_{16}H_{18}N_2OS$
Penicillin

$C_{16}H_{19}N_3O_5S$
Amoxicillin

$C_{17}H_{20}N_4O_6$
Riboflavin

$C_{19}H_{19}N_7O_6$
Folic Acid

$C_{19}H_{28}O_2$
Testosterone

$C_{20}H_{30}O$
Retinol

$C_{27}H_{45}OH$
Cholesterol

$C_{28}H_{34}N_2O_3$
Denatonium Benzoate

$C_{29}H_{50}O$
Alpha-Tocopherol

$C_{35}H_{28}O_5N_4Mg$
Chlorophyll

$C_{35}H_{30}O_5N_4Mg$
Chlorophyll

$C_{40}H_{56}$
Beta-Carotene

$C_{54}H_{70}O_6N_4Mg$
Chlorophyll

$C_{55}H_{70}O_6N_4Mg$
Chlorophyll

$C_{55}H_{72}O_5N_4Mg$
Chlorophyll

$C_{63}H_{88}CoN_{14}O_{14}P$
Cyanocobalamin

$C_{76}H_{52}O_{46}$
Tannic Acid

$CaCO_3$
Calcium Carbonate

$CaHPO_4$
Calcium Phosphate

CaO
Calcium Oxide

$Ca(OH)_2$
Calcium Hydroxide

$CaSO_4$
Calcium Sulfate

$CaSiO_3$
Calcium Silicate

$Ca(H_2PO_4)_2$
Calcium Phosphate

$Ca_3(PO_4)_2$
Calcium Phosphate

$(ClC_6H_4)_2CHCCl_3$
Dichlorodiphenyltrichloro-
ethane

$-ClO_4$
Perchlorates

CuO
Copper(II) Oxide

$CuSO_4$
Copper(II) Sulfate

Cu_2O
Copper(I) Oxide

FeO
Iron(II) Oxide

Fe_2O_3
Iron(III) Oxide

$HCHO$
Formaldehyde

HCl
Hydrogen Chloride

HNO_3
Nitric Acid

HOH
Water

HOOCCH$_2$C(OH)(COOH)CH$_2$COOH
  Citric Acid

HOOCCOOH
  Oxalic Acid

H$_2$O
  Water

H$_2$O$_2$
  Hydrogen Peroxide

H$_2$SO$_4$
  Sulfuric Acid

H$_3$BO$_3$
  Boric Acid

H$_3$PO$_4$
  Phosphoric Acid

HgS
  Mercury(II) Sulfide

KAl(SO$_4$)$_2$
  Aluminum Potassium
    Sulfate

KCl
  Potassium Chloride

KF
  Potassium Fluoride

KHCO$_3$
  Potassium Bicarbonate

KHC$_4$H$_4$O$_6$
  Potassium Bitartrate

KHSO$_4$
  Potassium Bisulfate

KI
  Potassium Iodide

KNO$_3$
  Potassium Nitrate

KOH
  Potassium Hydroxide

K$_2$CO$_3$ K$_{34}$NO$_{15}$
  Potassium Carbonate

K$_2$SO$_4$
  Potassium Sulfate

MgCl$_2$
  Magnesium Chloride

MgO
  Magnesium Oxide

Mg(OH)$_2$
  Magnesium Hydroxide

MgSO$_4$
  Magnesium Sulfate

Mg$_3$Si$_4$O$_{10}$(OH)$_2$
  Magnesium Silicate
    Hydroxide

(NH$_2$)$_2$CO
  Urea

NH$_3$
  Ammonia

(NH$_4$)$_2$SO$_4$
  Ammonium Sulfate

NH$_4$Cl
  Ammonium Chloride

NH$_4$NO$_3$
  Ammonium Nitrate

NH$_4$OH
  Ammonium Hydroxide

NO
  Nitric Oxide

NO$_2$
  Nitrogen Dioxide

N$_2$O
  Nitrous Oxide

NaBO$_3$
  Sodium Perborate

NaC$_2$H$_3$O$_2$
  Sodium Acetate

NaCl
  Sodium Chloride

NaClO
  Sodium Hypochlorite

NaF
  Sodium Fluoride

NaHCO$_3$
  Sodium Bicarbonate

NaH$_2$PO$_4$
  Sodium Phosphate

NaOH
  Sodium Hydroxide

Na$_2$B$_4$O$_7$
  Sodium Tetraborate

Na$_2$B$_4$O$_7$·10H$_2$O
  Sodium Tetraborate

Na$_2$CO$_3$
  Sodium Carbonate

Na$_2$HPO$_4$
  Sodium Phosphate

Na$_2$SO$_3$
  Sodium Sulfite

Na$_2$S$_2$O$_3$
  Sodium Thiosulfate

Na$_2$SiO$_3$
  Sodium Silicate

Na$_3$PO$_4$
  Sodium Phosphate

SO$_2$
  Sulfur Dioxide

SiO$_2$
  Silicon Dioxide

SnF$_2$
  Stannous Fluoride

ZnO
  Zinc Oxide

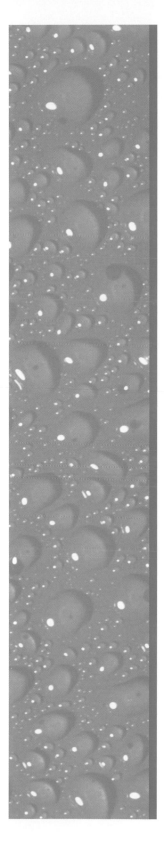

Compounds by Element

Dichlorodiphenyltrichloro-
  ethane
Dimethyl Ketone
Ethyl Acetate
Ethyl Alcohol
Ethylbenzene
Ethylene
Ethylene Glycol
Ethylene Oxide
Folic Acid
Formaldehyde
Fructose
Gamma-1,2,3,4,5,6-Hexachloro-
  cyclohexane
Gelatin
Glucose
Glycerol
Hexane
Isoamyl Acetate
Isoprene
Isopropyl Alcohol
Lactic Acid
Lactose
L-Aspartyl-L-Phenylalanine
  Methyl Ester
Luminol
Menthol
Methane
Methyl Alcohol
Methyl Mercaptan
Methyl-t-butyl Ether
Monosodium Glutamate
N,N-Diethyl-3-Methyl-
  benzamide
Naphthalene
Naproxen
Niacin
Nicotine
Nitroglycerin
Nylon 6 and Nylon 66
Oxalic Acid
Pectin
Penicillin
Petrolatum
Petroleum
Phenol
Poly(Styrene-Butadiene-
  Styrene)

Polycarbonates
Polyethylene
Polymethyl Methacrylate
Polypropylene
Polysiloxane
Polystyrene
Polytetrafluoroethylene
Polyurethane
Polyvinyl Chloride
Potassium Bicarbonate
Potassium Bitartrate
Potassium Carbonate
Propane
Propylene
Pyridoxine
Retinol
Riboflavin
Saccharin
Sodium Acetate
Sodium Bicarbonate
Sodium Carbonate
Sodium Cyclamate
Sodium Polyacrylate
Styrene
Sucrose
Sucrose Polyester
Tannic Acid
Testosterone
Theobromine
Thiamine
Toluene
Triclocarban
Triclosan
Urea
Vanillin

CHLORINE

2,2'-Dichlorodiethyl Sulfide
Ammonium Chloride
Carbon Tetrachloride
Chloroform
Dichlorodifluoromethane
Dichlorodiphenyltrichloro-
  ethane
Gamma-1,2,3,4,5,6-Hexachloro-
  cyclohexane
Hydrogen Chloride

Magnesium Chloride
Perchlorates
Polyvinyl Chloride
Potassium Chloride
Sodium Chloride
Sodium Hypochlorite
Thiamine
Triclocarban
Triclosan

COBALT

Cyanocobalamin

COPPER

Copper(I) Oxide
Copper(II) Oxide
Copper(II) Sulfate

FLUORINE

Aluminum Fluoride
Dichlorodifluoromethane
Polytetrafluoroethylene
Potassium Fluoride
Sodium Fluoride
Stannous Fluoride

HYDROGEN

1,3-Butadiene
2-(4-Isobutylphenyl)propionic
  Acid
2,2'-Dichlorodiethyl Sulfide
2,4,6-Trinitrotoluene
Acetaminophen
Acetic acid
Acetylene
Acetylsalicylic acid
Alpha-Tocopherol
Ammonia
Ammonium Chloride
Ammonium Hydroxide
Ammonium Nitrate
Ammonium Sulfate
Amoxicillin
Amyl Acetate
Amyl Nitrite
Ascorbic Acid

Benzene
Benzoic Acid
Beta-Carotene
Boric Acid
Butane
Butyl Acetate
Butyl Mercaptan
Butylated Hydroxyanisole and
  Butylated Hydroxytoluene
Caffeine
Calcium Hydroxide
Calcium Phosphate
Camphor
Cellulose
Cellulose Nitrate
Cellulose Xanthate
Chloroform
Chlorophyll
Cholesterol
Cinnamaldehyde
Citric Acid
Collagen
Cumene
Cyanoacrylate
Cyanocobalamin
Denatonium Benzoate
Dichlorodiphenyltrichloro-
  ethane
Dimethyl Ketone
Ethyl Acetate
Ethyl Alcohol
Ethylbenzene
Ethylene
Ethylene Glycol
Ethylene Oxide
Folic Acid
Formaldehyde
Fructose
Gamma-1,2,3,4,5,6-Hexachloro-
  cyclohexane
Gelatin
Glucose
Glycerol
Hexane
Hydrogen Chloride
Isoamyl Acetate
Isoprene
Isopropyl Alcohol

Lactic Acid
Lactose
L-Aspartyl-L-Phenylalanine
  Methyl Ester
Luminol
Magnesium Hydroxide
Magnesium Silicate
  Hydroxide
Menthol
Methane
Methyl Alcohol
Methyl Mercaptan
Methyl-t-butyl Ether
Monosodium Glutamate
N,N-Diethyl-3-Methyl-
  benzamide
Naphthalene
Naproxen
Niacin
Nicotine
Nitric Acid
Nitroglycerin
Nylon 6 and Nylon 66
Oxalic Acid
Pectin
Penicillin
Petrolatum
Petroleum
Phenol
Phosphoric Acid
Poly(Styrene-Butadiene-
  Styrene)
Polycarbonates
Polyethylene
Polymethyl
  Methacrylate
Polypropylene
Polysiloxane
Polystyrene
Polyurethane
Polyvinyl Chloride
Potassium Bicarbonate
Potassium Bisulfate
Potassium Bitartrate
Potassium Hydroxide
Propane
Propylene
Pyridoxine

Retinol
Riboflavin
Saccharin
Sodium Acetate
Sodium Bicarbonate
Sodium Cyclamate
Sodium Hydroxide
Sodium Polyacrylate
Styrene
Sucrose
Sucrose Polyester
Sulfuric Acid
Tannic Acid
Testosterone
Theobromine
Thiamine
Toluene
Triclocarban
Triclosan
Urea
Vanillin
Water

IODINE

Potassium Iodide
Silver Iodide

IRON

Iron(II) Oxide
Iron(III) Oxide

MAGNESIUM

Chlorophyll
Magnesium Chloride
Magnesium Hydroxide
Magnesium Oxide
Magnesium Silicate
  Hydroxide
Magnesium Sulfate

MERCURY

Mercury(II) Sulfide

NITROGEN

2,4,6-Trinitrotoluene
Acetaminophen

Ammonia
Ammonium Chloride
Ammonium Hydroxide
Ammonium Nitrate
Ammonium Sulfate
Amoxicillin
Amyl Nitrite
Caffeine
Cellulose Nitrate
Chlorophyll
Collagen
Cyanoacrylate
Cyanocobalamin
Denatonium Benzoate
Folic Acid
Gelatin
L-Aspartyl-L-Phenylalanine
    Methyl Ester
Luminol
Monosodium Glutamate
N,N-Diethyl-3-Methyl-
    benzamide
Niacin
Nicotine
Nitric Acid
Nitric Oxide
Nitrogen Dioxide
Nitroglycerin
Nylon 6 and Nylon 66
Penicillin
Polyurethane
Potassium Nitrate
Pyridoxine
Riboflavin
Saccharin
Silver Nitrate
Sodium Cyclamate
Theobromine
Thiamine
Triclocarban
Urea
Nitrous Oxide

OXYGEN

2-(4-Isobutylphenyl)propionic
    Acid
2,4,6-Trinitrotoluene
Acetaminophen

Acetic acid
Acetylsalicylic acid
Alpha-Tocopherol
Aluminum Hydroxide
Aluminum Potassium Sulfate
Aluminum Oxide
Ammonium Hydroxide
Ammonium Nitrate
Ammonium Sulfate
Amoxicillin
Amyl Acetate
Amyl Nitrite
Ascorbic Acid
Benzoic Acid
Boric Acid
Butyl Acetate
Butylated Hydroxyanisole
    and Butylated Hydro-
    xytoluene
Caffeine
Calcium Carbonate
Calcium Hydroxide
Calcium Oxide
Calcium Phosphate
Calcium Silicate
Calcium Sulfate
Camphor
Carbon Dioxide
Carbon Monoxide
Cellulose
Cellulose Nitrate
Cellulose Xanthate
Chlorophyll
Cholesterol
Cinnamaldehyde
Citric Acid
Collagen
Copper(I) Oxide
Copper(II) Oxide
Copper(II) Sulfate
Cyanoacrylate
Cyanocobalamin
Denatonium Benzoate
Dimethyl Ketone
Ethyl Acetate
Ethyl Alcohol
Ethylene Glycol
Ethylene Oxide

Folic Acid
Formaldehyde
Fructose
Gelatin
Glucose
Glycerol
Hydrogen Peroxide
Iron(II) Oxide
Iron(III) Oxide
Isoamyl Acetate
Isopropyl Alcohol
Lactic Acid
Lactose
L-Aspartyl-L-Phenylalanine
    Methyl Ester
Luminol
Magnesium Hydroxide
Magnesium Oxide
Magnesium Silicate
    Hydroxide
Magnesium Sulfate
Menthol
Methyl Alcohol
Methyl-t-butyl Ether
Monosodium Glutamate
N,N-Diethyl-3-Methyl-
    benzamide
Naproxen
Niacin
Nitric Acid
Nitric Oxide
Nitrogen Dioxide
Nitroglycerin
Nitrous Oxide
Nylon 6 and Nylon 66
Oxalic Acid
Pectin
Penicillin
Perchlorates
Petroleum
Phenol
Phosphoric Acid
Polycarbonates
Polymethyl Methacrylate
Polysiloxane
Polyurethane
Potassium Bicarbonate
Potassium Bisulfate

Potassium Bitartrate
Potassium Carbonate
Potassium Hydroxide
Potassium Nitrate
Potassium Sulfate
Pyridoxine
Retinol
Riboflavin
Saccharin
Silicon Dioxide
Silver Nitrate
Silver(I) Oxide
Sodium Acetate
Sodium Bicarbonate
Sodium Carbonate
Sodium Cyclamate
Sodium Hydroxide
Sodium Hypochlorite
Sodium Perborate
Sodium Phosphate
Sodium Polyacrylate
Sodium Silicate
Sodium Sulfite
Sodium Tetraborate
Sodium Thiosulfate
Sucrose
Sucrose Polyester
Sulfur Dioxide
Sulfuric Acid
Tannic Acid
Testosterone
Theobromine
Thiamine
Triclocarban
Triclosan
Urea
Vanillin
Water
Zinc Oxide

PHOSPHORUS

Calcium Phosphate

Phosphoric Acid
Sodium Phosphate

POTASSIUM

Aluminum Potassium
  Sulfate
Potassium Bicarbonate
Potassium Bisulfate
Potassium Bitartrate
Potassium Carbonate
Potassium Chloride
Potassium Fluoride
Potassium Hydroxide
Potassium Iodide
Potassium Nitrate
Potassium Sulfate

SILICON

Calcium Silicate
Magnesium Silicate
  Hydroxide
Polysiloxane
Silicon Dioxide
Sodium Silicate

SILVER

Silver Iodide
Silver Nitrate
Silver(I) Oxide
Silver(I) Sulfide

SODIUM

Cellulose Xanthate
Monosodium Glutamate
Sodium Acetate
Sodium Bicarbonate
Sodium Carbonate
Sodium Chloride
Sodium Fluoride
Sodium Hydroxide

Sodium Hypochlorite
Sodium Perborate
Sodium Phosphate
Sodium Polyacrylate
Sodium Silicate
Sodium Sulfite
Sodium Tetraborate
Sodium Thiosulfate

SULFUR

2,2'-Dichlorodiethyl
  Sulfide
Aluminum Potassium
  Sulfate
Ammonium Sulfate
Amoxicillin
Butyl Mercaptan
Calcium Sulfate
Cellulose Xanthate
Copper(II) Sulfate
Magnesium Sulfate
Mercury(II) Sulfide
Methyl Mercaptan
Penicillin
Potassium Bisulfate
Potassium Sulfate
Saccharin
Silver(I) Sulfide
Sodium Cyclamate
Sodium Sulfite
Sodium Thiosulfate
Sulfur Dioxide
Sulfuric Acid
Thiamine

TIN

Stannous Fluoride

ZINC

Zinc Oxide

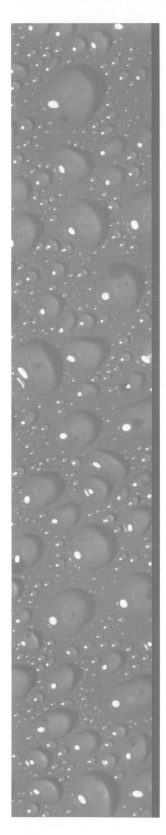

Compounds by Type

ACID

2-(4-Isobutylphenyl)propionic
  Acid
Acetic acid
Acetylsalicylic acid
Ascorbic Acid
Benzoic Acid
Boric Acid
Butyl Acetate
Citric Acid
Denatonium Benzoate
Folic Acid
Hydrogen Chloride
Lactic Acid
Naproxen
Niacin
Nitric Acid
Oxalic Acid
Penicillin
Phosphoric Acid
Potassium Bicarbonate
Potassium Bisulfate
Potassium Bitartrate
Sodium Bicarbonate
Sulfuric Acid
Tannic Acid

ALCOHOL

Ethyl Alcohol
Ethylene Glycol
Glycerol
Isopropyl Alcohol
Menthol
Methyl Alcohol
Retinol

ALDEHYDE

Cinnamaldehyde
Formaldehyde

ALKALOID

Caffeine
Nicotine
Theobromine

ALKANE

Butane
Hexane
Methane
Propane

ALKENE

1,3-Butadiene
Ethylene
Propylene

ALKYNE

Acetylene

AMIDE

Acetaminophen

BASE

Aluminum Hydroxide
Ammonia
Ammonium Hydroxide
Caffeine
Calcium Hydroxide
Magnesium Hydroxide
Potassium Hydroxide
Sodium Hydroxide
Theobromine

CARBOHYDRATE

Cellulose
Cellulose Nitrate
Fructose
Glucose
Lactose
Sucrose

CARBOXYLIC ACID

Acetic acid
Acetylsalicylic acid
Butyl Acetate
Citric Acid
Lactic Acid
Naproxen
Niacin
Oxalic Acid

ESTER

Amyl Acetate
Amyl Nitrite
Cyanoacrylate
Ethyl Acetate
Isoamyl Acetate
L-Aspartyl-L-Phenylalanine
   Methyl Ester
Nitroglycerin

ETHER

Ethylene Oxide
Methyl-t-butyl Ether
Vanillin

HYDROCARBON

1,3-Butadiene
Acetylene
Benzene
Beta-Carotene
Butane
Cumene
Ethylbenzene
Ethylene
Hexane
Isoprene
Methane
Naphthalene
Propane
Propylene
Styrene
Toluene

INORGANIC

Aluminum Fluoride
Aluminum Hydroxide
Aluminum Oxide
Aluminum Potassium Sulfate
Ammonia
Ammonium Chloride
Ammonium Hydroxide
Ammonium Nitrate
Ammonium Sulfate
Boric Acid
Calcium Carbonate
Calcium Hydroxide
Calcium Oxide
Calcium Phosphate
Calcium Silicate
Calcium Sulfate
Carbon Dioxide
Carbon Monoxide
Copper(I) Oxide
Copper(II) Oxide
Copper(II) Sulfate
Hydrogen Chloride
Iron(II) Oxide
Iron(III) Oxide
Magnesium Chloride
Magnesium Hydroxide
Magnesium Oxide

Magnesium Silicate Hydroxide
Magnesium Sulfate
Mercury(II) Sulfide
Nitric Acid
Nitric Oxide
Nitrogen Dioxide
Nitrous Oxide
Perchlorates
Phosphoric Acid
Polysiloxane
Potassium Bicarbonate
Potassium Bisulfate
Potassium Bitartrate
Potassium Carbonate
Potassium Chloride
Potassium Fluoride
Potassium Hydroxide
Potassium Iodide
Potassium Nitrate
Potassium Sulfate
Silicon Dioxide
Silver Iodide
Silver Nitrate
Silver(I) Oxide
Silver(I) Sulfide
Sodium Acetate
Sodium Bicarbonate
Sodium Carbonate
Sodium Chloride
Sodium Fluoride
Sodium Hydroxide
Sodium Hypochlorite
Sodium Perborate
Sodium Phosphate
Sodium Silicate
Sodium Sulfite
Sodium Tetraborate
Sodium Thiosulfate
Stannous Fluoride
Sulfur Dioxide
Sulfuric Acid
Water
Zinc Oxide

KETONE

Camphor
Dimethyl Ketone

METALLIC OXIDE

Aluminum Oxide
Calcium Oxide
Copper(I) Oxide
Copper(II) Oxide
Iron(II) Oxide
Iron(III) Oxide
Magnesium Oxide
Silver(I) Oxide
Zinc Oxide

NONMETALLIC OXIDE

Carbon Dioxide
Carbon Monoxide
Hydrogen Peroxide
Nitric Oxide
Nitrogen Dioxide
Nitrous Oxide
Silicon Dioxide
Sulfur Dioxide

ORGANIC

1,3-Butadiene
2-(4-Isobutylphenyl)propionic
   Acid
2,2'-Dichlorodiethyl Sulfide
2,4,6-Trinitrotoluene
Acetaminophen
Acetic acid
Acetylene
Acetylsalicylic acid
Alpha-Tocopherol
Amoxicillin
Amyl Acetate
Amyl Nitrite
Ascorbic Acid
Benzene
Benzoic Acid
Beta-Carotene
Butane
Butyl Acetate
Butyl Mercaptan
Butylated Hydroxyanisole and
   Butylated Hydroxytoluene
Caffeine
Camphor

Carbon Tetrachloride
Cellulose
Cellulose Nitrate
Cellulose Xanthate
Chloroform
Chlorophyll
Cholesterol
Cinnamaldehyde
Citric Acid
Collagen
Cumene
Cyanoacrylate
Cyanocobalamin
Denatonium
   Benzoate
Dichlorodifluoromethane
Dichlorodiphenyltrichloro-
   ethane
Dimethyl Ketone
Ethyl Acetate
Ethyl Alcohol
Ethylbenzene
Ethylene
Ethylene Glycol
Ethylene Oxide
Folic Acid
Formaldehyde
Fructose
Gamma-1,2,3,4,5,6-Hexachloro-
   cyclohexane
Glucose
Glycerol
Hexane
Hydrogen Peroxide
Isoamyl Acetate
Isoprene
Isopropyl Alcohol
Lactic Acid
Lactose
L-Aspartyl-L-Phenylalanine
   Methyl Ester
Luminol
Menthol
Methane
Methyl Alcohol
Methyl Mercaptan
Methyl-t-butyl Ether
Monosodium Glutamate

N,N-Diethyl-3-Methylbenza-
   mide
Naphthalene
Naproxen
Niacin
Nicotine
Nitroglycerin
Nylon 6 and Nylon 66
Oxalic Acid
Penicillin
Phenol
Poly(Styrene-Butadiene-
   Styrene)
Polycarbonates
Polyethylene
Polymethyl Methacrylate
Polypropylene
Polystyrene
Polytetrafluoroethylene
Polyurethane
Polyvinyl Chloride
Propane
Propylene
Pyridoxine
Retinol
Riboflavin
Saccharin
Sodium Cyclamate
Sodium Polyacrylate
Styrene
Sucrose
Sucrose Polyester
Tannic Acid
Testosterone
Theobromine
Thiamine
Toluene
Triclocarban
Triclosan
Urea
Vanillin

PHENOL

Butylated Hydroxyanisole
   and Butylated Hydroxy-
   toluene
Phenol

POLYMER

Cellulose
Cellulose Nitrate
Cellulose Xanthate
Nylon 6 and Nylon 66
Poly(Styrene-Butadiene-
   Styrene)
Polycarbonates
Polyethylene
Polymethyl Methacrylate
Polypropylene
Polysiloxane
Polystyrene
Polytetrafluoroethylene
Polyurethane
Polyvinyl Chloride
Sodium Polyacrylate

SALT

Aluminum Fluoride
Aluminum Potassium Sulfate
Ammonium Chloride
Ammonium Nitrate
Ammonium Sulfate
Calcium Carbonate
Calcium Phosphate
Calcium Silicate
Calcium Sulfate
Copper(II) Sulfate
Magnesium Chloride
Magnesium Silicate
   Hydroxide
Magnesium Sulfate
Mercury(II) Sulfide
Monosodium Glutamate
Perchlorates
Potassium Bicarbonate
Potassium Bisulfate
Potassium Bitartrate
Potassium Carbonate
Potassium Chloride
Potassium Fluoride
Potassium Iodide
Potassium Nitrate
Potassium Sulfate
Silver Iodide
Silver Nitrate
Silver(I) Sulfide
Sodium Acetate
Sodium Bicarbonate
Sodium Carbonate
Sodium Chloride
Sodium Cyclamate
Sodium Fluoride
Sodium Hypochlorite
Sodium Perborate
Sodium Phosphate
Sodium Silicate
Sodium Sulfite
Sodium Tetraborate
Sodium Thiosulfate
Stannous Fluoride

VITAMIN

Alpha-Tocopherol
Ascorbic Acid
Cyanocobalamin
Folic Acid
Niacin
Pyridoxine
Retinol
Riboflavin
Thiamine

for further information

BOOKS

Attenborough, David. *The Private Life of Plants.* Princeton, NJ: Princeton University Press, 1995.

Brody, Tom. *Nutritional Biochemistry.* San Diego: Academic Press, 1998.

Buchanan, B. B., W. Gruissem, and R. L. Jones. *Biochemistry and Molecular Biology of Plants.* Rockville, MD: American Society of Plant Physiologists, 2000.

Buechel, K. H., et al. *Industrial Inorganic Chemistry.* New York: VCH, 2000.

Buschmann, Helmut, et al. *Analgesics: From Chemistry and Pharmacology to Clinical Application.* New York: Wiley-VCH, 2002.

Butler, A. R., and R. Nicholson. *Life, Death and Nitric Oxide.* London: Royal Society of Chemistry, 2003.

Cagin, Seth, and Philip Dray. *Between Earth and Sky: How CFCs Changed Our World and Endangered the Ozone Layer.* New York: Pantheon Books, 1993.

Carpenter, Kenneth J. *The History of Scurvy and Vitamin C.* Cambridge, UK: Cambridge University Press, 1986.

Carson, Rachel. *Silent Spring.* Boston: Houghton Mifflin, 1962.

Cavitch, Susan Miller. *The Natural Soap Book: Making Herbal and Vegetable-Based Soaps*. Markham, Canada: Storey Publishing, 1995.

Challem, Jack, and Melissa Diane Smith. *Basic Health Publications User's Guide to Vitamin E: Don't Be a Dummy: Become an Expert on What Vitamin E Can Do for Your Health*. North Bergen, NJ: Basic Health Publications, 2002.

Chalmers, Louis. *Household and Industrial Chemical Specialties*. Vol. 1. New York: Chemical Publishing Co., Inc., 1978.

Cherniske, Stephen. *Caffeine Blues: Wake Up to the Hidden Dangers of America's #1 Drug*. New York: Warner Books, 1998.

"Cholesterol, Other Lipids, and Lipoproteins." In *In-Depth Report*. Edited by Julia Goldrosen. Atlanta: A.D.A.M., 2004.

Cooper, P. W., and S. R. Kurowski. *Introduction to the Technology of Explosives*. New York: Wiley-VCH, 1997.

Cornell, Rochelle M., and Udo Schwertmann. *The Iron Oxides: Structure, Properties, Reactions, and Uses*, 2nd ed. New York: Wiley-VCH, 2003.

CRC Handbook of Chemistry and Physics. David R. Lide, editor in chief. 86th ed. Boca Raton, FL: Taylor & Francis, 2005.

Dean, Carolyn. *The Miracle of Magnesium*. New York: Ballantine Books, 2003.

Dick, John S., and R. A. Annicelli, eds. *Rubber Technology: Compounding and Testing for Performance*. Cincinnati, OH: Hanser Gardner Publications, 2001.

Dunlap, Thomas. *DDT: Scientists, Citizens, and Public Policy*. Princeton, NJ: Princeton University Press, 1983.

Dwyer, Bob, et al. *Carbon Monoxide: A Clear and Present Danger*. Mount Prospect, IL: ESCO Press, 2004.

Eades, Mary Dan. *The Doctor's Complete Guide to Vitamins and Minerals*. New York: Dell, 2000.

Environment Canada, Health Canada. *Ethylene Oxide*. Ottawa: Environment Canada, 2001.

Food Antioxidants: Technological, Toxicological, and Health Perspectives, D. L. Madhavi, S. S. Deshpande, and D. K. Salunkhe, eds. New York: Dekker, 1996.

Gahlinger, Paul M. *Illegal Drugs: A Complete Guide to Their History*. Salt Lake City, UT: Sagebrush Press, 2001.

Genge, Ngaire E. *The Forensic Casebook: The Science of Crime Scene Investigation*. New York: Ballantine, 2002.

Grimm, Tom, and Michele Grimm. *The Basic Book of Photography*. New York: Plume Books, 2003.

Harte, John, et al. *Toxics A to Z.* Berkeley: University of California Press, 1991.

Hermes, Matthew E. *Enough for One Lifetime: Wallace Carothers, Inventor of Nylon.* Philadelphia: Chemical Heritage Foundation, 1996.

Holgate, S. T., et al. *Air Pollution and Health.* New York: Academic Press, 1999.

Hyne, Norman J. *Nontechnical Guide to Petroleum Geology, Exploration, Drilling and Production,* 2nd ed. Tulsa, OK: Pennwell Books, 2001.

Jeffreys, Diarmuid. *Aspirin: The Remarkable Story of a Wonder Drug.* New York: Bloomsbury, 2004.

Johnson, Peter S. *Rubber Processing: An Introduction.* Cincinnati, OH: Hanser Gardner Publications, 2001.

Karger-Kocsis, J. *Polypropylene—An A-Z Reference.* New York: Springer-Verlag, 1998.

Kirk-Othmer Encyclopedia of Chemical Technology, 4th ed. New York: John Wiley & Sons, 1991.

Knox, J. Paul, and Graham B. Seymour, eds. *Pectins and Their Manipulation.* Boca Raton, FL: CFC Press, June 2002.

Kurlansky, Mark. *Salt: A World History.* New York: Walker, 2002.

Leffler, William L. *Petroleum Refining in Nontechnical Language.* Tulsa, OK: Pennwell Books, 2000.

Mead, Clifford, and Thomas Hager, eds. *Linus Pauling: Scientist and Peacemaker.* Portland, OR: Oregon State University Press, 2001.

Mebane, Robert C., and Thomas R. Rybolt. *Plastics and Polymers.* New York: Twenty-First Century, 1995.

Meikle, Jeffrey L. *American Plastic: A Cultural History.* New Brunswick, NJ: Rutgers University Press, 1995.

Menhard, Francha Roffe. *The Facts about Inhalants.* New York: Benchmark Books, 2004.

Misra, Chanakya. *Industrial Alumina Chemicals.* Washington, DC: American Chemical Society, 1986.

Mosby's Medical, Nursing, and Allied Health Dictionary, 5th ed. St. Louis: Mosby, 1998.

Multhauf, Robert P., and Christine M. Roane. "Nitrates." In *Dictionary of American History.* 3rd ed., vol. 6. Stanley I. Kutler, ed. New York: Charles Scribner's Sons, 2003.

Nabors, Lyn O'Brien, ed. *Alternative Sweeteners (Food Science and Technology),* 3rd rev. London: Marcel Dekker, 2001.

Packer, Lester, and Carol Colman. *The Antioxidant Miracle: Put Lipoic Acid, Pycogenol, and Vitamins E and C to Work for You.* New York: Wiley, 1999.

Patnaik, Pradyot. *Handbook of Inorganic Chemicals.* New York: McGraw-Hill, 2003.

Rain, Patricia. *Vanilla: A Cultural History of the World's Most Popular Flavor and Fragrance.* Edited by Jeremy P. Tarcher. New York: Penguin Group USA, 2004.

Rainsford, K. D., ed. *Ibuprofen: A Critical Bibliographic Review.* Bethesda, MD: CCR Press, 1999.

Richardson, H. W., ed. *Handbook of Copper Compounds and Applications.* New York: Marcel Dekker, 1997.

Sherman, Josepha, and Steve Brick. *Fossil Fuel Power.* Mankato, MN: Capstone Press, 2003.

Snyder, C. H. *The Extraordinary Chemistry of Ordinary Things*, 4th ed. New York: John Wiley and Sons, 2002.

Strange, Veronica. *The Meaning of Water.* Oxford, UK: Berg Publishers, 2004.

Stratmann, Linda. *Chloroform: The Quest for Oblivion.* Phoenix Mill, UK: Sutton Publishing Co., 2003.

Tegethoff, F. Wolfgana, with Johannes Rohleder and Evelyn Kroker, eds. *Calcium Carbonate: From the Cretaceous Period into the 21st Century.* Boston: Birkhäuser Verlag, 2001.

Tyman, J. H. P. *Synthetic and Natural Phenols.* Amsterdam: Elsevier, 1996.

Ware, George W. *The Pesticide Book.* Batavia, IL: Mesiter, 1999.

Weinberg, Alan Bennet, and Bonnie K. Bealer. *The World of Caffeine: The Science and Culture of the World's Most Popular Drug.* New York: Routledge, 2002.

Weissermel, Klaus, and Hans-Jürgen Arpe. *Industrial Organic Chemistry.* Weinheim, Germany: Wiley-VCH, 2003, 117-122.

Wyman, Carolyn. *JELL-O: A Biography.* Fort Washington, PA: Harvest Books, 2001.

PERIODICALS

"Another Old-Fashioned Product Vindicates Itself." *Medical Update* (October 1992): 6.

Arnst, Catherine. "A Preemptive Strike against Cancer." *Business Week* (June 7, 2004): 48.

Baker, Linda. "The Hole in the Sky (Ozone Layer)." *E* (November 2000): 34.

Bauman, Richard. "Getting Skunked: Understanding the Antics behind the Smell." *Backpacker* (May 1993): 30-31.

Drake, Geoff. "The Lactate Shuttle—Contrary to What You've Heard, Lactic Acid Is Your Friend." *Bicycling* (August 1992): 36.

Fox, Berry. "Not Fade Away." *New Scientist* (March 1, 2003); 40.

"Global Ethyl, Butyl Acetate Demand Expected to Rebound." *The Oil and Gas Journal* (April 24, 2000): 27.

Gorman, Christine. "The Bomb Lurking in the Garden Shed." *Time* (May 1, 1995): 54.

Karaim, Reed. "Not So Fast with the DDT: Rachel Carson's Warnings Still Apply." *American Scholar* (June 2005): 53-60.

Keenan, Faith. "Blocking Liver Damage." *Business Week* (October 21, 2002): 147.

Kluger, Jeffrey. "The Buzz on Caffeine." *Time* (December 20, 2004): 52.

Lazear, N. R. "Polycarbonate: High-Performance Resin." *Advanced Materials & Processes* (February 1995): 43-45.

Legwold, Gary. "Hydration Breakthrough." *Bicycling* (July 1994): 72-73.

Liu, Guanghua. "Chinese Cinnabar." *The Mineralogical Record* (January-February 2005): 69-80.

Malakoff, David. "Public Health: Aluminum Is Put on Trial as a Vaccine Booster." *Science* (May 26, 2000): 1323.

Mardis, Anne L. "Current Knowledge of the Health Effects of Sugar Intake." *Family Economics and Nutrition Review* (Winter 2001): 88-91.

McGinn, Anne Platt. "Malaria, Mosquitoes, and DDT: The Toxic War against a Global Disease." *World Watch* (May 1, 2002): 10-17.

Metcalfe, Ed, et al. "Sweet Talking." *The Ecologist* (June 2000): 16.

Milius, Susan. "Termites Use Mothballs in Their Nests." *Science News* (May 2, 1998): 228.

Neff, Natalie. "No Laughing Matter." *Auto Week* (May 19, 2003): 30.

"Nitroglycerin: Dynamite for the Heart." *Chemistry Review* (November 1999): 28.

O'Neil, John. "And It Doesn't Taste Bad, Either." *New York Times* (November 30, 2004): F9.

Pae, Peter. "Sobering Side of Laughing Gas." *Washington Post* (September 16, 1994): B1.

Palvetz, Barry A. "A Bowl of Hope, Bucket of Hype?" *The Scientist* (April 2, 2001): 15.

Rawls, Rebecca. "Nitroglycerin Explained." *Chemical & Engineering News* (June 10, 2002): 12.

Rowley, Brian. "Fizzle or Sizzle? Potassium Bicarbonate Could Help Spare Muscle and Bone." *Muscle & Fitness* (December 2002): 72.

Russell, Justin. "Fuel of the Forgotten Deaths." *New Scientist* (February 6, 1993): 21-23.

Schramm, Daniel. "The North American USP Petrolatum Industry." *Soap & Cosmetics*, (January 2002): 60-63.

Stanley, Peter. "Nitric Oxide." *Biological Sciences Review* (April 2002): 18-20.

"The Stink that Stays." *Popular Mechanics* (December 2004): 26.

Strobel, Warren P. "Saddam's Lingering Atrocity." *U.S. News & World Report* (November 27, 2000): 52.

"Strong Muscle and Bones." *Prevention* (June 1, 1995): 70-73.

"Taking Supplements of the Antioxidant." *Consumer Reports* (September 2003): 49.

Travis, J. "Cool Discovery: Menthol Triggers Cold-Sensing Protein." *Science News* (February 16, 2002): 101-102.

U.S. Department of Health and Human Services. "Methanol Toxicity." *American Family Physician* (January 1993): 163-171.

"Unusual Thermal Defence by a Honeybee against Mass Attack by Hornets." *Nature* (September 28, 1995): 334-336.

"USDA Approves Phosphate to Reduce Salmonella in Chicken." *Environmental Nutrition* (February 1993): 3.

Vartan, Starre. "Pretty in Plastic: Pleather Is a Versatile, though Controversial, Alternative to eather." *E* (September-October 2002): 53-54.

"Vitamins: The Quest for Just the Right Amount." *Harvard Health Letter* (June 2004): 1.

Walter, Patricia A. "Dental Hypersensitivity: A Review." *The Journal of Contemporary Dental Practice* (May 15, 2005): 107-117.

Young, Jay A. "Copper (II) Sulfate Pentahydrate." *Journal of Chemical Education* (February 2002): 158.

WEBSITES

"The A to Z of Materials." Sydney, Australia: Azom.com. http://www.azom.com/ (accessed on March 1, 2006).

"Air Toxics Website." U.S. Environmental Protection Agency Technology Transfer Network http://www.epa.gov/ttn/atw/ (accessed on March 10, 2006).

Calorie Control Council. Atlanta, GA: Calorie Control Council. http://www.caloriecontrol.org/

CHEC's HealtheHouse: The Resource for Environmental Health Risks Affecting Your Children. Los Angeles, CA: Children's Health Environmental Coalition. http://www.checnet.org/ ehouse (accessed on March 13, 2006).

Chemfinder.com. Cambridge, MA: CambridgeSoft Corporation. http://www.chemfinder.cambridgesoft.com (accessed on March 13, 2006).

"Chemical Backgrounders." Itasca, IL: National Safety Council. http://www.nsc.org/library/chemical/ (accessed on March 1, 2006).

"Chemical Profiles." Scorecard. The Pollution Information Site. Washington, DC: Green Media Toolshed. http://www.scorecard. org/chemical-profiles/ (accessed on March 1, 2006).

The Chemistry Store. Cayce, SC: ChemistryStore.com, Inc. http:// www.chemistrystore.com/ (accessed on March 13, 2006).

Chemistry.org. Washington, DC: American Chemical Society. http://www.chemistry.org/portal/a/c/s/1/home.html (accessed on March 13, 2006).

Cheresources.com: Online Chemical Engineering Information. Midlothian, VA: The Chemical Engineers' Resource Page. http://www.cheresources.com (accessed on March 13, 2006).

"Dietary Supplement Fact Sheets." Bethesda, MD: U.S. National Institutes of Health, Office of Dietary Supplements. http://ods.od.nih.gov/Health_Information/Information_About_ Individual_Dietary_Supplements.aspx (accessed on March 13, 2006).

Drugs.com. Auckland, New Zealand: Drugsite Trust. http:// www.drugs.com (accessed on March 13, 2006).

EnvironmentalChemistry.com. Portland, ME: Environmental Chemistry.com. http://environmentalchemistry.com (accessed on March 13, 2006).

Exploratorium: the museum of science, art and human perception. San Francisco, CA: Exploratorium at the Palace of Fine Arts. http://www.exploratorium.edu/ (accessed on March 13, 2006).

ExToxNet: The Extension Toxicology Network. Corvallis, OR: Oregon State University. http://extoxnet.orst.edu/ (accessed on March 13, 2006).

Fibersource: The Manufactured Fibers Industry. Arlington, VA: Fiber Economics Bureau. http://www.fibersource.com (accessed on March 13, 2006).

General Chemistry Online. Frostburg, MD: Frostburg State University, Department of Chemistry. http://antoine.frostburg.edu/chem/senese/101/index.shtml (accessed on March 13, 2006).

Household Products Database. Bethesda, MD: U.S. National Library of Medicine. http://householdproducts.nlm.nih.gov/ (accessed on March 13, 2006).

Integrated Risk Information System. Washington, DC: U.S. Environmental Protection Agency. http://www.epa.gov/iris/index.html (accessed on March 13, 2006).

"International Chemical Safety Cards (ICSCs)." Geneva, Switzerland: International Labour Organization. http://www.ilo.org/public/english/protection/safework/cis/products/icsc/index.htm (accessed on March 1, 2006).

"International Chemical Safety Cards." Atlanta, GA: U.S. National Institute for Occupational Safety and Health. http://www.cdc.gov/niosh/ipcs/icstart.html (accessed on March 1, 2006).

IPCS INTOX Databank. Geneva, Switzerland: International Programme on Chemical Safety. http://www.intox.org/ (accessed on March 13, 2006).

Kimball's Biology Pages. Andover, MA: John W. Kimball. http://biology-pages.info (accessed on March 13, 2006).

Linus Pauling Institute Micronutrient Information Center. Corvallis, OR: Oregon State University, Linus Pauling Institute. http://lpi.oregonstate.edu/infocenter/ (accessed on March 13, 2006).

MadSci Network. Boston, MA: Third Sector New England. http://www.madsci.org (accessed on March 13, 2006).

Medline Plus. Bethesda, MD: U.S. National Library of Medicine. http://www.nlm.nih.gov/medlineplus (accessed on March 1, 2006).

Mineral Information Institute. Golden, CO: Mineral Information Institute. http://www.mii.org (accessed March 13, 2006).

Mineralogy Database. Spring, TX: Webmineral.com. http://webmineral.com/ (accessed on March 13, 2006).

"Molecule of the Month Page." Bristol, United Kingdom: University of Bristol School of Chemistry. http://www.chm.bris.ac.uk/motm/motm.htm (accessed on March 1, 2006).

National Historic Chemical Landmarks. Washington, DC: American Chemical Society. http://acswebcontent.acs.org/landmarks/index.html (accessed on March 13, 2006).

NIST Chemistry Webbook. Gaithersburg, MD: U.S. National Institute of Standards and Technology. http://webbook.nist.gov (accessed on March 13, 2006).

PAN Pesticides Database. San Francisco, CA: Pesticide Action Network, North America. http://www.pesticideinfo.org (accessed March 13, 2006).

"Polymer Science Learning Center." Hattiesburg, Mississippi: University of Southern Mississippi, Department of Polymer Science. http://www.pslc.ws/ (accessed on March 13, 2006).

PubChem. Bethesda, MD: U.S. National Library of Medicine. http://pubchem.ncbi.nlm.nih.gov/ (accessed on March 13, 2006).

Reciprocal.net. Bloomington, IN: Indiana University Molecular Structure Center. http://www.reciprocalnet.org/ (accessed on March 13, 2006).

The Science Center: A Teacher's Guide to Educational Resources on the Internet. Arlington, VA: Chlorine Chemistry Council. http://www.science-education.org (accessed on March 13, 2006).

Shakhashiri, Bassam Z. "Chemical of the Week." Science Is Fun. Madison, WI: University of Wisconsin. http://scifun.chem.wisc.edu/chemweek/chemweek.html (accessed on March 13, 2006).

3Dchem.com. Oxford, United Kingdom: University of Oxford, Department of Chemistry. http://www.3dchem.com (accessed on March 13, 2006).

"ToxFAQs™" Atlanta, GA: U.S. Agency for Toxic Substances and Disease Registry. http://www.atsdr.cdc.gov/toxfaq.html (accessed on March 13, 2006).

"Toxicological Profile Information Sheets." Atlanta, GA: U.S. Agency for Toxic Substances and Disease Registry. http://www.atsdr.cdc.gov/toxpro2.html (accessed on March 1, 2006).

TOXNET: Toxicology Data Network. Bethesda, MD: U.S. National Library of Medicine. http://toxnet.nlm.nih.gov (accessed on March 13, 2006).

World of Molecules Home Page. http://www.worldofmolecules.com/ (accessed on March 13, 2006).

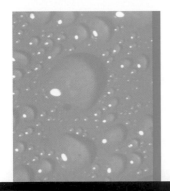

index

This index is sorted
word-by-word. Italic
type indicates volume
numbers; **boldface**
indicates main
entries; (ill.) indicates
illustrations.

A

C

D

G

Q

R

Vaccines, *1:*47

Vanillin, *3:***873-877**, 874 (ill.)

Vaseline, *2:*547-549

Vasodilators, *1:*89-90

Vermillion, *2:*440

Vicks VapoRub, *1:*173; *2:*437-438

Vinegar, *1:*23-26

Viscose Corp. of America, *1:*208

Vision, *3:*679

Vitamins

    alpha-tocopherol (E), *1:*37-40

    ascorbic acid (C), *1:*93-97

    beta-carotene, *1:*109-113; *3:*678, *3:*680

    cyanocobalamin (B$_{12}$), *1:*265-268

    folic acid (B$_9$), *2:*321-324

    niacin (B$_3$), *2:*483-486

    pyridoxine (B$_6$), *3:*673-676

    retinol (A), *3:*677-681

    riboflavin (B$_2$), *3:*683-687

    thiamine (B$_1$), *3:*822, 847-851

    See also Minerals (nutritional)

Volpenhein, Robert, *3:*813

Vonnegut, Bernard, *3:*703

Wackenroder, Heinrich Wilhelm
   Ferdinand, *1:*110

Waldie, David, *1:*211

Warning odors, *1:*87

Washing soda, *3:*729-733

Water, *3:***879-884**, 880 (ill.)

Water treatment
   aluminum potassium sulfate, *1:*55

ammonium sulfate, *1:*78
   fluoride, *1:*43; *3:*748-749
   sodium sulfite in, *3:*786
   softeners, *3:*737

Wedgwood, Thomas, *3:*706

Wehmer, Carl, *1:*235

Wells, Horace, *2:*514

Whipple, George Hoyt, *1:*266

Whitehall Laboratories, *1:*11

Wiegleb, Johann Christian, *2:*525

Wieland, Heinrich Otto, *1:*224-225

Wilbrand, Joseph, *1:*15

Williams, Robert R., *3:*849-850

Willow bark, *1:*31-32

Wills, Lucy, *2:*321

Winhaus, Adolf, *1:*224-225

Wislicenus, Johannes, *2:*392

Wöhler, Friedrich, *1:*27; *2:*525; *3:*867

Wood alcohol, *2:*449-453

Wood preservatives, *1:*248, 252

Woodward, Robert Burns, *1:*219, 225, 267

Wounds, sealing of, *1:*260-261

Wulff process, *1:*28

Wurtz, Charles Adolphe, *2:*313, 317

Xylene, *3:*855

Zeidler, Othmar, *1:*283

Zeolites, *1:*257

Ziegler, Karl, *2:*384, 580, *2:*582, 588-589

Zinc oxide, *3:***885-888**, 886 (ill.)